# SOIL TESTING LABORATORY MANUAL

Ming Xiao
*California State University, Fresno*

Kendall Hunt
publishing company

**Kendall Hunt**
publishing company

www.kendallhunt.com
*Send all inquiries to:*
4050 Westmark Drive
Dubuque, IA 52004-1840



Printed in the United States of America
10 9 8 7

ISBN 978-1-5249-2588-8

# Table of Contents

# PREFACE

This laboratory soil testing manual provides a detailed guide of fourteen common laboratory soil tests. The manual follows the ASTM standards published in the 2006 annual book. This manual can be used as a textbook in civil engineering undergraduate program as well as a reference in the geotechnical engineering practice. Each laboratory testing includes a brief introduction of the theoretical background, the applicable ASTM references, the types of soils and other materials needed, specimen preparation, the apparatus and tools needed, test procedure, and detailed data collection process. Each laboratory testing is explained using sufficient photo illustrations and includes detailed sample test results.

The author thanks the support of the Lyles College of Engineering at California State University, Fresno. Appreciation also extends to Stephanie Moffett and Wendy Pauly of Kendall Hunt Publishing, and to Jon Fuller, Publisher of Bent Tree Press, for their suggestions and help in the development of this manual. Clinton Jones, Juan Orozco, Eric Stewards, and Ian Bushell from ELE International provided prompt help on soil testing equipment. The author thanks the assistance and suggestions provided by Nathan Shwiyhat, staff engineer of BSK Associates (Fresno, CA), and Jose Gomez, graduate student at CSU Fresno.

Ming Xiao
Fresno, California

March 23, 2011

# TEST 1

# Water (Moisture) Content of Soil by Mass

## 1. Introduction

The objective of this test is to determine the moisture content of soil by mass. Usually the moisture in soil is water, so the moisture content is also commonly referred to as water content. The definition of water content is

$$w = \frac{M_w}{M_s} \times 100\%$$

where:

$w$ = water content, expressed in percentage,
$M_w$ = mass of water in the soil,
$M_s$ = dry soil mass = mass of soil solids.

## 2. References

- ASTM D 2216-05 "Standard Test Methods for Laboratory Determination of Water (Moisture) Content of Soil and Rock by Mass," *2006 Annual Book of ASTM Standards*, Volume 04.08, Soil and Rock (I). ASTM International.
- ASTM D 4643-00 "Standard Test Methods for Laboratory Determination of Water (Moisture) Content of Soil by the Microwave Oven Heating," *2006 Annual Book of ASTM Standards*, Volume 04.08, Soil and Rock (I). ASTM International.
- ASTM D 4959-00 "Standard Test Methods for Laboratory Determination of Water (Moisture) Content of Soil by Direct Heating," *2006 Annual Book of ASTM Standards*, Volume 04.08, Soil and Rock (I). ASTM International.

## 3. Soils and Other Materials

- Any type of soil

## 4. Apparatus

- Balance with readability of 0.1 g when the specimen mass is larger than 200 g, or readability of 0.01 g when the specimen mass is less than 200 g
- Drying oven maintaining a constant temperature of 110±5°C
- Thermometer of 0.1°C readability
- Specimen container with lid (2)

## 5. Preparation of Test Specimen

- Moist soil is used

## 6. Test Procedure

(1) Measure the mass of a dry and clean specimen container with lid.

Figure 1-1. Drying oven for water content measurement.

(2) Select a representative test specimen, place the moist soil in the container, and close the lid.
(3) Measure the mass of the container with lid and the soil.
(4) Remove the lid and place the container with soil in the drying oven; set the oven temperature at 110±5°C. Dry the soil overnight (12 to 16 hours).

***Note:***
- Gloves and safety goggles are needed for protection.
- If more rapid results are desired, drying the soil using a microwave oven or direct heating (gas stove, hotplate, hair drier, etc.) can be used per ASTM D4643 and D4959, respectively. Slightly less accurate results may be derived.
- If organic soil (such as peat) is tested, it is preferable to dry the materials at 60°C or in a desiccator at room temperature.

(5) Remove the container using gloves and measure the mass of the container (with lid closed) and the dry soil specimen. The lid is used to prevent the soil from absorbing moisture in the air.
(6) Calculate the moisture content.
(7) Repeat the above procedure to obtain a second moisture content value, and calculate the average value.
(8) Dispose of the dry soil and clean the containers.

## 7. Data Collection and Calculation

Table 1-1. Water Content Measurement

| | **Test 1** | **Test 2** |
|---|---|---|
| Mass of container with lid, $M_1$ (g) | | |
| Mass of container with lid and ***moist*** soil specimen, $M_2$ (g) | | |
| Mass of container with lid and ***dry*** soil specimen, $M_3$ (g) | | |
| Mass of water, $M_w = M_2 - M_3$ (g) | | |
| Mass of dry soil, $M_s = M_3 - M_1$ (g) | | |
| Moisture content, $w = \frac{M_w}{M_s} \times 100\%$ (%) | | |
| Average moisture content (%) | | |

## 8. Results and Analyses

Complete the calculation of Table 1-1 and report the average moisture content.

## SAMPLE DATA COLLECTION, CALCULATION, RESULTS, AND ANALYSES

| | **Test 1** | **Test 2** |
|---|---|---|
| Mass of container with lid, $M_1$ (g) | 31.22 | 31.43 |
| Mass of container with lid and ***moist*** soil specimen, $M_2$ (g) | 130.81 | 139.82 |
| Mass of container with lid and ***dry*** soil specimen, $M_3$ (g) | 120.13 | 127.74 |
| Mass of water, $M_w = M_2 - M_3$ (g) | 10.68 | 12.08 |
| Mass of dry soil, $M_s = M_3 - M_1$ (g) | 88.91 | 96.31 |
| Moisture content, $w = \frac{M_w}{M_s} \times 100\%$ (%) | 12.0 | 12.6 |
| Average moisture content (%) | 12.3 | |

The average moisture content is 12.3%.

Name ______________________________

# Test 1: Water (Moisture) Content of Soil by Mass

Question 1: What is the typical temperature in an oven for moisture content measurement of inorganic soils?

Question 2: A field soil sample has a wet mass of 500.0 g. After oven drying over night, the dry soil sample weighs 400.0 g. What is the moisture content of the field soil sample?

TEST 2

# Particle Size Distribution of Soils – Sieve Analysis

## 1. Introduction

The objective of this test is to measure the grain size distribution of sandy soils so that the soil can be classified. The measurement is conducted using a stack of sieves with different opening sizes, as indicated in Figure 2-1 and Table 2-1. The sieves are mounted on a mechanical sieve shaker to shake the soil, so that soil particles of different sizes are retained on sieves with corresponding openings.

Figure 2-1. U.S. standard sieves.

Figure 2-2. Sieve shaker.

Table 2-1. U.S. Standard Sieves

| U.S. Sieve | Size (mm) | U.S. Sieve | Size (mm) | U.S. Sieve | Size (mm) |
|---|---|---|---|---|---|
| 3-in | 75 | No. 10 | 2.00 | No. 50 | 0.300 |
| 2-in | 50 | No. 12 | 1.70 | No. 60 | 0.250 |
| 1.5 -in | 37.5 | No. 14 | 1.40 | No. 70 | 0.212 |
| 1-in | 25.0 | No. 16 | 1.18 | No. 80 | 0.180 |
| 3/4-in | 19.0 | No. 18 | 1.00 | No. 100 | 0.150 |
| 3/8-in | 9.5 | No. 20 | 0.85 | No. 120 | 0.125 |
| No. 4 | 4.75 | No. 25 | 0.71 | No. 140 | 0.106 |
| No. 5 | 4.00 | No. 30 | 0.60 | No. 170 | 0.088 |
| No. 6 | 3.35 | No. 35 | 0.500 | No. 200 | 0.075 |
| No. 7 | 2.80 | No. 40 | 0.425 | No. 270 | 0.053 |
| No. 8 | 2.36 | No. 45 | 0.355 | No. 325 | 0.044 |

## 2. References

- ASTM D 421 "Practice for Dry Preparation of Soil Samples for Particle-Size Analysis and Determination of Soil Constants," *2006 Annual Book of ASTM Standards*, Volume 04.08, Soil and Rock (I). ASTM International.
- ASTM D 422-63 (Reapproved 2002) "Standard Test Method for Particle-Size Analysis of Soils," *2006 Annual Book of ASTM Standards*, Volume 04.08, Soil and Rock (I). ASTM International.

## 3. Soils and Other Materials

- Granular soil

## 4. Apparatus

- Balance sensitive to 0.1 g
- A series of sieves (Table 2-1 and Figure 2-1)
- Mechanical sieve shaker (Figure 2-2)

## 5. Preparation of Test Specimen

Per ASTM D422, sieve analysis is conducted on soils retained on the No. 10 sieve and again on soils retained on the No. 200 sieve after the hydrometer test. For teaching purposes, granular soil can be used, although the procedure illustrated here applies to all types of soils.

For each test, 600 g of air-dried soil is needed.

## 6. Test Procedure

(1) Collect and weigh 600.0 g of air-dried sandy soil.
(2) Prepare a stack of sieves. The following sieves are recommended: No. 4 (4.75 mm), No. 10 (2.0 mm), No. 40 (0.425 mm), and No. 200 (0.075 mm). Other sieves may be added. The sieves should be in dry and clean condition.
(3) Weigh each sieve and the bottom pan and record the mass.
(4) Stack the sieves and the bottom pan together. Sieves with larger openings are at the top.
(5) Place the weighed soil onto the top sieve, and place the cover on top of the stack of sieves.
(6) Secure the stack of sieves on the sieve shaker.
(7) Turn on the sieve shaker and run it for 10 minutes. The sieve shaker should be attended while it is running.
(8) Stop the sieve shaker and remove the stack of sieves from the shaker.
(9) Carefully take off each sieve (starting from the top) and weigh the total mass of the sieve and soil retained on the sieve. Then weigh the mass of the pan and the soil retained on the pan.
(10) Restack the sieves and return them to the shaker to run for 1 minute. Repeat Steps (8) and (9). If the mass difference is more than 1% of the soil mass on any sieve, repeat this step until the mass difference of each sieve with soil is less than 1%.
(11) Record and calculate the soil mass retained on each sieve and pan. Dispose of the soil and clean the sieves.

## 7. Data Collection and Calculation

Table 2-2. Sieve Analysis

| **Sieve No.** | **Sieve Size (mm)** | **Sieve Mass (g)** | **Mass of Sieve and Soil (g)** | **Mass of Soil Retained (g)** | **Cumulative Soil Mass Retained (g)** | **Mass of Soil Passing the Sieve (g)** | **Percent Passing (%)** |
|---|---|---|---|---|---|---|---|
| | | | | | | | |
| | | | | | | | |
| | | | | | | | |
| | | | | | | | |
| | | | | | | | |
| | | | | | | | |
| | | | | | | | |

Total mass of soil used =
Total mass of soil recovered =
Mass loss of soil during sieve analysis =
Mass loss percentage of soil = (should be less than 2%)

## 8. Results and Analyses

(1) Complete the calculation of Table 2-2.

(2) Plot the grain size distribution in terms of particle sizes (same as sieve openings) vs. passing percentage of total soil mass on each sieve. The particle size is plotted in logarithmic scale, decreasing from left to right; the passing percentage is in arithmetic scale.

(3) From the curve, identify the particle diameter corresponding to the passing percentage of 10% (denoted as $D_{10}$), 30% (denoted as $D_{30}$), and 60% (denoted as $D_{60}$).

(4) Calculate the uniformity coefficient, $C_u$:

$$C_u = \frac{D_{60}}{D_{10}}$$

and the coefficient of curvature (also referred to as coefficient of gradation), $C_c$:

$$C_c = \frac{D_{30}^2}{D_{10} \cdot D_{60}}$$

(5) Determine whether the sandy soil is well or poorly graded, using the criteria in Table 2-3.

Table 2-3. Criteria of Soil Gradation

| | **Well Graded** | **Poorly Graded** |
|---|---|---|
| **Sand** | $C_u \geq 6$ AND $1 \leq C_c \leq 3$ | $C_u < 6$ OR $C_c$ is not in the range of 1 and 3 inclusive. |
| **Gravel** | $C_u \geq 4$ AND $1 \leq C_c \leq 3$ | $C_u < 4$ OR $C_c$ is not in the range of 1 and 3 inclusive. |

## SAMPLE DATA COLLECTION, RESULTS, AND ANALYSES

Table 2-4. Sieve Analysis

| Sieve No. | Sieve Size (mm) | Sieve Mass (g) | Mass of Sieve and Soil (g) | Mass of Soil Retained (g) | Cumulative Soil Mass Retained (g) | Mass of Soil Passing the Sieve (g) | Percent Passing (%) |
|---|---|---|---|---|---|---|---|
| 4 | **4.75** | 522.0 | 522.0 | 0 | 0 | 598.9 | **100** |
| 10 | **2.00** | 440.6 | 450.0 | 9.4 | 9.4 | 589.5 | **98** |
| 20 | **0.85** | 415.5 | 477.4 | 61.9 | 71.3 | 527.6 | **88** |
| 40 | **0.425** | 394.0 | 712.9 | 318.9 | 390.2 | 208.7 | **35** |
| 100 | **0.150** | 345.2 | 531.6 | 186.4 | 576.6 | 22.3 | **4** |
| 200 | **0.075** | 339.2 | 353.9 | 14.7 | 591.3 | 7.6 | **1** |
| Pan | N/A | 287.4 | 295.0 | 7.6 | 598.9 | - | - |

Total mass of soil used = 600.0 g
Total mass of soil recovered = 598.9 g
Mass loss of soil during sieve analysis = 1.1 g
Mass loss percentage of soil = 0.2% (should be less than 2%)

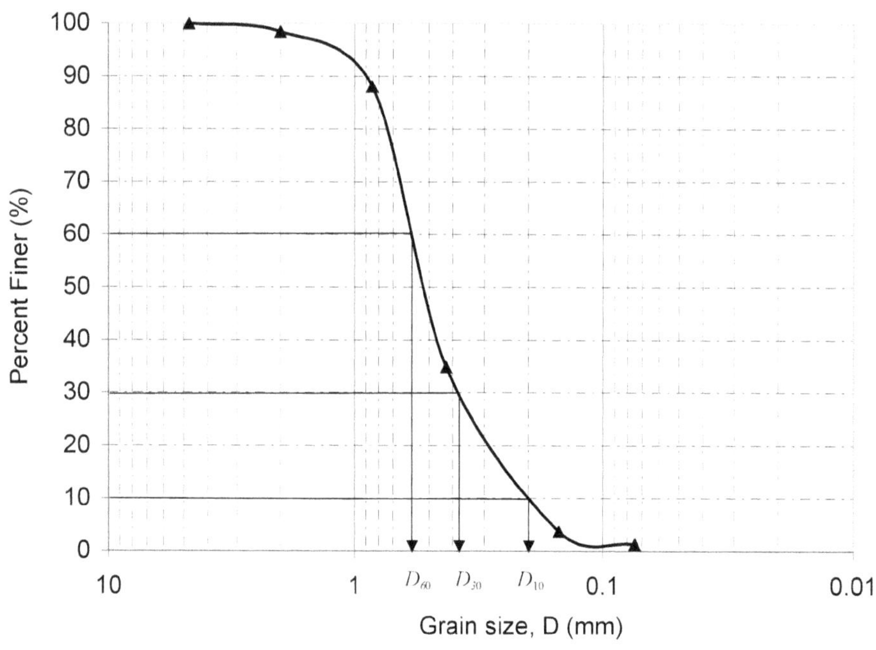


Figure 2-3. Grain size distribution of the tested sandy soil.

From the graph, determine $D_{10}$ = 0.20 mm, $D_{30}$ = 0.39 mm, and $D_{60}$ = 0.60 mm.

The coefficient of uniformity, $C_u = \frac{D_{60}}{D_{10}} = \frac{0.60}{0.20} = 3.0 < 6$

The coefficient of curvature, $C_c = \frac{D_{30}^{\ 2}}{D_{10} \cdot D_{60}} = \frac{0.39^2}{0.20 \times 0.60} = 1.27$

Therefore, the sandy soil is poorly graded.

Name ______________________________

# Test 2: Particle Size Distribution of Soils – Sieve Analysis

Question 1: What type of soil is the sieve analysis typically used for?

Question 2: What are the four sieves that are typically used in a sieve analysis? List their U.S. sieve numbers and sieve openings.

Question 3: How long should a sieve shaker be run in a sieve analysis?

# TEST 3

# Particle Size Distribution of Soils – Hydrometer Test

## 1. Introduction

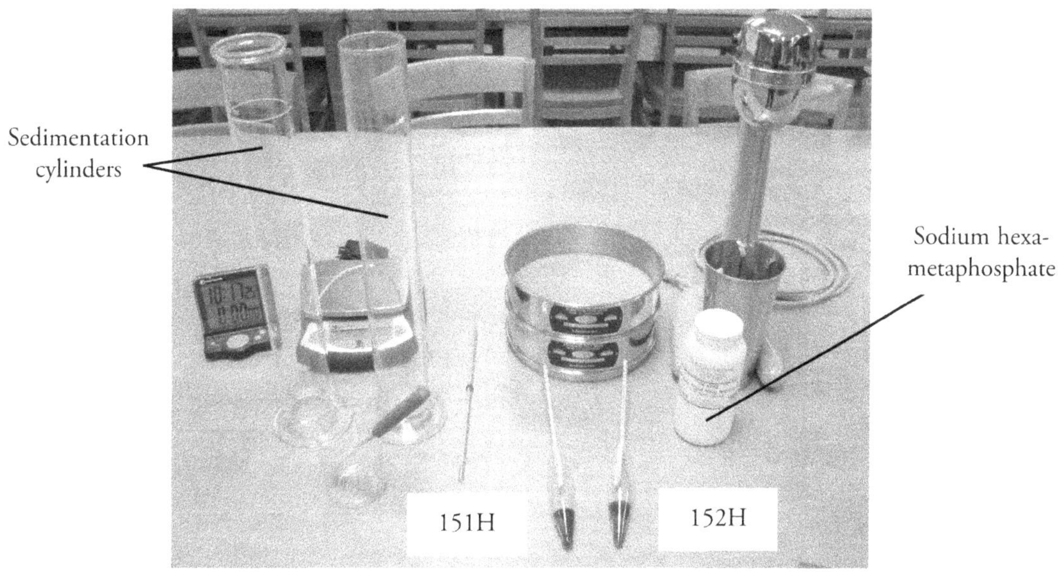


Figure 3-1. Apparatus for hydrometer test.

The objective of this test is to measure the particle (grain) size distribution of fine-grained soil so that the soil can be classified.

The hydrometer test is used to obtain the particle size distribution of fine soils by measuring the density of fine soil suspensions at different times. Two types of hydrometers can be used in the test: 151H and 152H (Figure 3-1). Since the particle suspensions settle with time, the fluid density changes. The particle size can be calculated according to Stokes' law:

$$D = \sqrt{\frac{30\eta}{980(G_{Soil} - G_{suspension})} \times \frac{L}{t}} = K\sqrt{\frac{L}{t}} \qquad (1)$$

where:

| | | |
|---|---|---|
| $D$ (mm) | = | particle diameter, |
| $\eta$ (g/(cm·s)) | = | dynamic viscosity of the suspending medium and is approximated as the viscosity of water, $1.002\times10^{-2}$ g/(cm·s) at 20°C, |
| $G_{soil}$ | = | specific gravity of the soil particles, |

$G_{suspension}$ = specific gravity of the suspension; it can be approximated as the value of water, 1.0,

$L$ (cm) = effective length, which is the distance from the surface of the suspension to the level at which the density of the suspension is measured; $L$ is obtained from Table 3-1 based on the hydrometer reading (corrected for meniscus; see explanation below) and depending on the type of the hydrometer,

$t$ (min) = time interval from the beginning of the sedimentation to the taking of the reading.

In Eq. (1), the lumped parameter, $K$, can be obtained from Table 3-2.

A dispersing agent (sodium hexametaphosphate, 40 g/L) is added to the soil suspension to prevent clay particle flocculation. This causes the specific gravity of the resulting liquid to be appreciably greater than that of distilled water. The hydrometer reading should be corrected to account for the effects of specific gravity of the liquid with dispersing agent, temperature, and meniscus:

(1) ***Correction of meniscus, $C_m$:*** Since it is not possible to read the bottom of the meniscus due to the opaque suspension, readings can only be taken at the top of meniscus and a correction should be applied. Insert a hydrometer into clear distilled water. The reading difference between the top and the bottom of the meniscus is $C_m$.

(2) ***Correction of temperature, $C_T$:*** The hydrometer and the suspension are calibrated at 20°C. Inaccuracy at other temperatures should be corrected.

(3) ***Correction of specific gravity, $C_{Gs}$:*** In pure distilled water, a hydrometer should have a reading of 1.0 for 151H or 0 for 152H; with the dispersing agent, a hydrometer has a different reading, which is $C_{Gs}$.

The three corrections can be lumped into a composite correction, $C_{composite}$:

$$C_{composite} = C_{Gs} - C_m - C_T \tag{2}$$

and $C_{composite}$ can be experimentally determined, using the following method:

Prepare 1,000 mL distilled water with 5.0 g of dispersing agent in a glass cylinder at a constant temperature. Insert the hydrometer and wait for a short period to permit the hydrometer to equalize the temperature of the liquid. Read the hydrometer at the top of the meniscus as $R$. For the 151H hydrometer, $C_{composite} = R - 1$; for the 152H hydrometer, $C_{composite} = R$.

The corrected hydrometer reading is

$$R_C = R_{act} - C_{composite} \tag{3}$$

where:

$R_{act}$ = the actual hydrometer reading.

The percentage ($P$) of soil mass remaining in suspension, i.e., smaller than diameter $D$, is calculated using the corrected hydrometer reading ($R_C$) with the three corrections applied.

- For hydrometer 151H:

$$P = \frac{G_{soil}}{G_{soil} - G_{suspension}} \times (R_C - G_{suspension}) \times \frac{100{,}000}{W} \qquad (4)$$

where:

$W$ (g) = total oven-dried soil mass used,
$G_{\text{suspension}}$ = specific gravity of the suspension; use $G_{\text{suspension}}$ = 1.0.

- For hydrometer 152H:

$$P = \frac{R_C \cdot a}{W} \times 100\% \qquad (5)$$

where:

$a$ = correction factor based on the specific gravity of soil particles (Table 3-3).

## 2. References


- ASTM D 421 "Practice for Dry Preparation of Soil Samples for Particle-Size Analysis and Determination of Soil Constants," *2006 Annual Book of ASTM Standards*, Volume 04.08, Soil and Rock (I). ASTM International.
- ASTM D 422-63 (Reapproved 2002) "Standard Test Method for Particle-Size Analysis of Soils," *2006 Annual Book of ASTM Standards*, Volume 04.08, Soil and Rock (I). ASTM International.
- ASTM D 1140-00 "Standard Test Methods for Amount of Materials in Soils Finer than the No. 200 (75-µm) Sieve," *2006 Annual Book of ASTM Standards*, Volume 04.08, Soil and Rock (I). ASTM International.


## 3. Soils and Other Materials

- Air-dried soils passing the No. 10 (2.00 mm) sieve: for each test, 65 g if the soil is mostly clay and silt, or 115 g if the soil is mostly sand
- Dispersing agent: sodium hexametaphosphate (40 g per test)
- Distilled or demineralized water (about 2.5 L per test)

## 4. Apparatus

- No. 10 sieve
- Balance sensitive to 0.01 g
- Stirring apparatus (high-speed mechanical stirrer or air dispersion)
- Hydrometer (151H or 152H)
- Two sedimentation cylinders with one rubber stopper: a glass cylinder 18 inches (457 mm) in height and 2.5 inches (63.5 mm) in diameter, marked for a volume of 1,000 mL
- Thermometer accurate to 1°F (0.5°C)
- Two beakers of 250 mL capacity
- Timer
- Water content container

Table 3-1. Values of Effective Depth, $L$, Based on Hydrometer Reading

| Hydrometer 151H | | Hydrometer 152H | | | |
|---|---|---|---|---|---|
| Actual Hydrometer Reading | Effective Depth, $L$ (cm) | Actual Hydrometer Reading | Effective Depth, $L$ (cm) | Actual Hydrometer Reading | Effective Depth, $L$ (cm) |
| 1.000 | 16.3 | 0 | 16.3 | 31 | 11.2 |
| 1.001 | 16.0 | 1 | 16.1 | 32 | 11.1 |
| 1.002 | 15.8 | 2 | 16.0 | 33 | 10.9 |
| 1.003 | 15.5 | 3 | 15.8 | 34 | 10.7 |
| 1.004 | 15.2 | 4 | 15.6 | 35 | 10.6 |
| 1.005 | 15.0 | 5 | 15.5 | 36 | 10.4 |
| 1.006 | 14.7 | 6 | 15.3 | 37 | 10.2 |
| 1.007 | 14.4 | 7 | 15.2 | 38 | 10.1 |
| 1.008 | 14.2 | 8 | 15.0 | 39 | 9.9 |
| 1.009 | 13.9 | 9 | 14.8 | 40 | 9.7 |
| 1.010 | 13.7 | 10 | 14.7 | 41 | 9.6 |
| 1.011 | 13.4 | 11 | 14.5 | 42 | 9.4 |
| 1.012 | 13.1 | 12 | 14.3 | 43 | 9.2 |
| 1.013 | 12.9 | 13 | 14.2 | 44 | 9.1 |
| 1.014 | 12.6 | 14 | 14.0 | 45 | 8.9 |
| 1.015 | 12.3 | 15 | 13.8 | 46 | 8.8 |
| 1.016 | 12.1 | 16 | 13.7 | 47 | 8.6 |
| 1.017 | 11.8 | 17 | 13.5 | 48 | 8.4 |
| 1.018 | 11.5 | 18 | 13.3 | 49 | 8.3 |
| 1.019 | 11.3 | 19 | 13.2 | 50 | 8.1 |
| 1.020 | 11.0 | 20 | 13.0 | 51 | 7.9 |
| 1.021 | 10.7 | 21 | 12.9 | 52 | 7.8 |
| 1.022 | 10.5 | 22 | 12.7 | 53 | 7.6 |
| 1.023 | 10.2 | 23 | 12.5 | 54 | 7.4 |
| 1.024 | 10.0 | 24 | 12.4 | 55 | 7.3 |
| 1.025 | 9.7 | 25 | 12.2 | 56 | 7.1 |
| 1.026 | 9.4 | 26 | 12.0 | 57 | 7.0 |
| 1.027 | 9.2 | 27 | 11.9 | 58 | 6.8 |
| 1.028 | 8.9 | 28 | 11.7 | 59 | 6.6 |
| 1.029 | 8.6 | 29 | 11.5 | 60 | 6.5 |
| 1.030 | 8.4 | 30 | 11.4 | | |
| 1.031 | 8.1 | | | | |
| 1.032 | 7.8 | | | | |
| 1.033 | 7.6 | | | | |
| 1.034 | 7.3 | | | | |
| 1.035 | 7.0 | | | | |
| 1.036 | 6.8 | | | | |
| 1.037 | 6.5 | | | | |
| 1.038 | 6.2 | | | | |

**Table 3-2. Values of *K* in Eq. (1)**

| Temperature (°C) | Specific Gravity of Soil Particles | | | | | | | | |
|---|---|---|---|---|---|---|---|---|---|
| | 2.45 | 2.50 | 2.55 | 2.60 | 2.65 | 2.70 | 2.75 | 2.80 | 2.85 |
| 16 | 0.01530 | 0.01505 | 0.01481 | 0.01457 | 0.01435 | 0.01414 | 0.01394 | 0.01374 | 0.01356 |
| 17 | 0.01511 | 0.01486 | 0.01462 | 0.01439 | 0.01417 | 0.01396 | 0.01376 | 0.01356 | 0.01338 |
| 18 | 0.01492 | 0.01467 | 0.01443 | 0.01421 | 0.01399 | 0.01378 | 0.01359 | 0.01339 | 0.01321 |
| 19 | 0.01474 | 0.01449 | 0.01425 | 0.01403 | 0.01382 | 0.01361 | 0.01342 | 0.01320 | 0.01305 |
| 20 | 0.01456 | 0.01431 | 0.01408 | 0.01386 | 0.01365 | 0.01344 | 0.01325 | 0.01307 | 0.01289 |
| 21 | 0.01438 | 0.01414 | 0.01391 | 0.01369 | 0.01348 | 0.01328 | 0.01309 | 0.01291 | 0.01273 |
| 22 | 0.01421 | 0.01397 | 0.01374 | 0.01353 | 0.01332 | 0.01312 | 0.01294 | 0.01276 | 0.01258 |
| 23 | 0.01404 | 0.01381 | 0.01358 | 0.01337 | 0.01317 | 0.01297 | 0.01279 | 0.01261 | 0.01243 |
| 24 | 0.01388 | 0.01365 | 0.01342 | 0.01321 | 0.01301 | 0.01282 | 0.01264 | 0.01246 | 0.01229 |
| 25 | 0.01372 | 0.01349 | 0.01327 | 0.01306 | 0.01286 | 0.01267 | 0.01249 | 0.01232 | 0.01215 |
| 26 | 0.01357 | 0.01334 | 0.01312 | 0.01291 | 0.01272 | 0.01253 | 0.01235 | 0.01218 | 0.01201 |
| 27 | 0.01342 | 0.01319 | 0.01297 | 0.01277 | 0.01258 | 0.01239 | 0.01221 | 0.01204 | 0.01188 |
| 28 | 0.01327 | 0.01304 | 0.01283 | 0.01264 | 0.01244 | 0.01255 | 0.01208 | 0.01191 | 0.01175 |
| 29 | 0.01312 | 0.01290 | 0.01269 | 0.01249 | 0.01230 | 0.01212 | 0.01195 | 0.01178 | 0.01162 |
| 30 | 0.01298 | 0.01276 | 0.01256 | 0.01236 | 0.01217 | 0.01199 | 0.01182 | 0.01165 | 0.01149 |

**Table 3-3. Values of Correction Factor, *a*, Used in Eq. (5)**

| Specific Gravity of Soil Particles | Correction Factor, *a* |
|---|---|
| 2.95 | 0.94 |
| 2.90 | 0.95 |
| 2.85 | 0.96 |
| 2.80 | 0.97 |
| 2.75 | 0.98 |
| 2.70 | 0.99 |
| 2.65 | 1.00 |
| 2.60 | 1.01 |
| 2.55 | 1.02 |
| 2.50 | 1.03 |
| 2.45 | 1.05 |

## 5. Preparation of Test Specimen

(1) Pass air-dried soil through the No. 10 (2.00 mm) sieve, and collect the soil passing the sieve. If the soil is mostly clay or silt, 50 g should be collected; if the soil is mostly sand, 100 g should be collected.

(2) Weigh 5.0 g of sodium hexametaphosphate powder, pour into a beaker, add 125 mL of distilled water, and stir until the powder dissolves. The dispersing agent concentration is 40 g/L.
(3) Place the soil in a second beaker and cover with 125 mL of the sodium hexametaphosphate solution. Stir the soil using a straightedge until it is thoroughly wetted. Wash all the solids on the straightedge into the beaker using distilled water (without dispersing agent). Cover the beaker and let it soak for at least **16 hours**.
(4) Measure the water content of the air-dried soil using the oven (refer to Test 1).

Figure 3-2. Soil slurry.

Figure 3-3. Hydrometer test in progress.

## 6. Test Procedure

(1) Set aside two glass sedimentation cylinders. Fill one with distilled water to the 1,000 mL mark. Insert the hydrometer into this cylinder and record the meniscus correction, $C_m$.
(2) Measure the composite correction, $C_{composite}$.
- Weigh 5 g of sodium hexametaphosphate powder, pour into a beaker, add 125 mL of distilled water, and stir until the powder dissolves. The dispersing agent concentration is 40 g/L.
- Transfer all the solution into the 1,000 mL cylinder and add 875 mL of distilled water to the 1,000 mL mark.
- Insert the hydrometer into the 1,000 mL cylinder. Wait for a short period to permit the hydrometer to equalize the temperature of the liquid. Read the hydrometer at the top of the meniscus. The reading is denoted as *R*. For the 151H hydrometer, $C_{composite} = R - 1$; for the 152H hydrometer, $C_{composite} = R$.
- Place the hydrometer back to the cylinder with distilled water. Dispose of the dispersing solution in the cylinder and rinse it with distilled water.

(3) At the end of the soaking period, disperse the sample further using a mechanical stirrer or air dispersion. In this test, the traditional mechanical stirrer is used. The air dispersion method is described in ASTM D-422.
   a. Transfer the soil-water slurry (Figure 3-2) from the beaker into the dispersion cup and use distilled water to wash any soil residue from the beaker into the cup.

b. Add distilled water if necessary to fill the cup more than half full.
c. Stir for 1 minute.

(4) Immediately after the dispersion, transfer the soil-water slurry to the empty glass sedimentation cylinder and use distilled water to wash any residue from the dispersion cup into the cylinder. Add distilled water until the total volume is 1,000 mL.

(5) Cap the cylinder with a rubber stopper and turn the cylinder upside down and back 60 times in 60 seconds. Then set the cylinder on a table and remove the stopper.
*Note:* The test should be conducted in a constant-temperature room; otherwise, the cylinder should be placed in a constant-temperature water bath.

(6) Insert the hydrometer into the cylinder to take hydrometer readings at **0.5, 1, 2, 5, 15, 30, 60, 250, and 1440 minutes** (Figure 3-3).
- Slowly insert the hydrometer about 20 seconds before the reading is due in order to equalize the temperature of the hydrometer and the suspension.
- Take the reading at the top of the meniscus.
- Immediately after the reading, carefully remove the hydrometer from the suspension and place it back in the cylinder with clean distilled water.
- After each reading, take the temperature of the suspension by inserting the thermometer into the suspension.

(7) After the final hydrometer reading, wash the soil in the cylinder through a No. 200 sieve. If required, a sieve analysis can be performed on the soils retained above the No. 200 sieve after oven drying per ASTM D1140 and D422.

## 7. Data Collection and Calculation

Hydrometer type:

Specific gravity of the soil particles, $G_s$ =

Water content of the air-dried soil, $w$ =

Mass of the air-dried soil, $W_1$ =

Dry soil mass used,

$$W = \frac{W_1}{1 + \frac{w(\%)}{100\%}} =$$

Temperature, $T$ =

Correction of meniscus, $C_m$ =

Composite correction, $C_{composite}$ =

Constant factor used in Eq. (1), $K$ =

Correction factor used in Eq. (5) if using 152H (from Table 3-3), $a$ =

| Time, $t$ (min) | Actual Hydrometer Reading, $R_{act}$ | Actual Hydrometer Reading Corrected for Meniscus, $R' = R_{act} + C_m$ | Effective Length, $L$ (from Table 1 using $R'$) (cm) | Particle Diameter, $D$ $D = K\sqrt{\frac{L}{t}}$ (mm) | Corrected Hydrometer Reading, $R_C = R_{act} - C_{composite}$ | Percent Finer, $P$ (Eq. 4 or 5) (%) |
|---|---|---|---|---|---|---|
| 0.5 | | | | | | |
| 1 | | | | | | |
| 2 | | | | | | |
| 5 | | | | | | |
| 15 | | | | | | |
| 30 | | | | | | |
| 60 | | | | | | |
| 250 | | | | | | |
| 1440 | | | | | | |

## 8. Results and Analyses

(1) Complete the calculation in the above table to find the particle diameters and percent passing.

(2) Plot the particle size distribution with the particle diameter as the abscissa (log scale) and the percent passing as the ordinate (arithmetic scale).

# SAMPLE DATA COLLECTION AND CALCULATION, RESULTS, AND ANALYSES

## (1). Sample 1 Using the 151H Hydrometer

Hydrometer type: 151H

Specific gravity of the soil particles, $G_s = 2.66$

Water content of the air-dried soil, $w = 3.3\%$

Mass of the air-dried soil, $W_1 = 64.6$

Dry soil mass used,

$$W = \frac{W_1}{1+\frac{w(\%)}{100\%}} = 62.5\text{g}$$

Temperature, $T = 20.0°\text{C}$

Correction of meniscus, $C_m = 0.001$

Composite correction, $C_{composite} = 1.004 - 1 = 0.004$

Constant factor used in Eq. (1), $K = 0.01365$

| Time, $t$ (min) | Actual Hydrometer Reading, $R_{act}$ | Actual Hydrometer Reading Corrected for Meniscus, $R' = R_{act}+C_m$ | Effective Length, $L$ (from Table 1 using $R'$) (cm) | Particle Diameter, $D$ $D=K\sqrt{\frac{L}{t}}$ (mm) | Corrected Hydrometer Reading, $R_C = R_{act} - C_{composite}$ | Percent Finer, $P$ (Eq. 4) (%) |
|---|---|---|---|---|---|---|
| 0.5 | 1.028 | 1.029 | 8.6 | 0.0566 | 1.024 | 61 |
| 1 | 1.027 | 1.028 | 8.9 | 0.0407 | 1.023 | 58 |
| 2 | 1.026 | 1.027 | 9.2 | 0.0293 | 1.022 | 56 |
| 4 | 1.023 | 1.024 | 10.0 | 0.0216 | 1.019 | 48 |
| 8 | 1.020 | 1.021 | 10.7 | 0.0158 | 1.016 | 40 |
| 15 | 1.018 | 1.019 | 11.3 | 0.0118 | 1.014 | 35 |
| 30 | 1.016 | 1.017 | 11.8 | 0.0086 | 1.012 | 30 |
| 60 | 1.014 | 1.015 | 12.3 | 0.0062 | 1.010 | 25 |
| 250 | 1.012 | 1.013 | 12.9 | 0.0031 | 1.008 | 20 |
| 1440 | 1.010 | 1.011 | 13.4 | 0.0013 | 1.006 | 15 |

Eq. 4: $$P(\%) = \frac{G_{soil}}{G_{soil}-1.0} \times (R_c - 1.0) \times \frac{100000}{W}$$

**(2). Sample 2 Using the 152H Hydrometer**

Hydrometer type: 152H
Specific gravity of the soil particles, $G_s$ = 2.66
Water content of the air-dried soil, $w$ = 3.3%
Mass of the air-dried soil, $W_1$ = 65.6g

Dry soil mass used, $W = \dfrac{W_1}{1+\dfrac{w(\%)}{100\%}}$ =63.5g

Temperature, $T$ = 20.0°C
Correction of meniscus, $C_m$ = 1.0
Composite correction, $C_{composite}$ = 8
Constant factor used in Eq. (1), $K$ = 0.01365
Correction factor used in Eq. (5) (from Table 3-3), $a$ = 1.00

| Time, $t$ (min) | Actual Hydrometer Reading, $R_{act}$ | Actual Hydrometer Reading Corrected for Meniscus, $R' = R_{act}+C_m$ | Effective Length, $L$ (from Table 1 using $R'$) (cm) | Particle Diameter, $D$ $D=K\sqrt{\frac{L}{t}}$ (mm) | Corrected Hydrometer Reading, $R_C = R_{act} - C_{composite}$ | Percent Finer, $P$ (Eq. 5) (%) |
|---|---|---|---|---|---|---|
| 0.5 | 44 | 45 | 8.9 | 0.0576 | 36 | 57 |
| 1 | 43 | 44 | 9.1 | 0.0412 | 35 | 55 |
| 2 | 41 | 42 | 9.4 | 0.0296 | 33 | 52 |
| 4 | 37 | 38 | 10.1 | 0.0217 | 29 | 46 |
| 8 | 33 | 34 | 10.7 | 0.0158 | 25 | 39 |
| 15 | 30 | 31 | 11.2 | 0.0118 | 22 | 35 |
| 30 | 28 | 29 | 11.5 | 0.0085 | 20 | 32 |
| 69 | 24 | 25 | 12.2 | 0.0057 | 16 | 25 |
| 253 | 20 | 21 | 12.9 | 0.0031 | 12 | 19 |
| 1440 | 16 | 17 | 13.5 | 0.0013 | 8 | 13 |

Eq. 5: $P = \dfrac{R_c \cdot a}{W} \times 100\%$

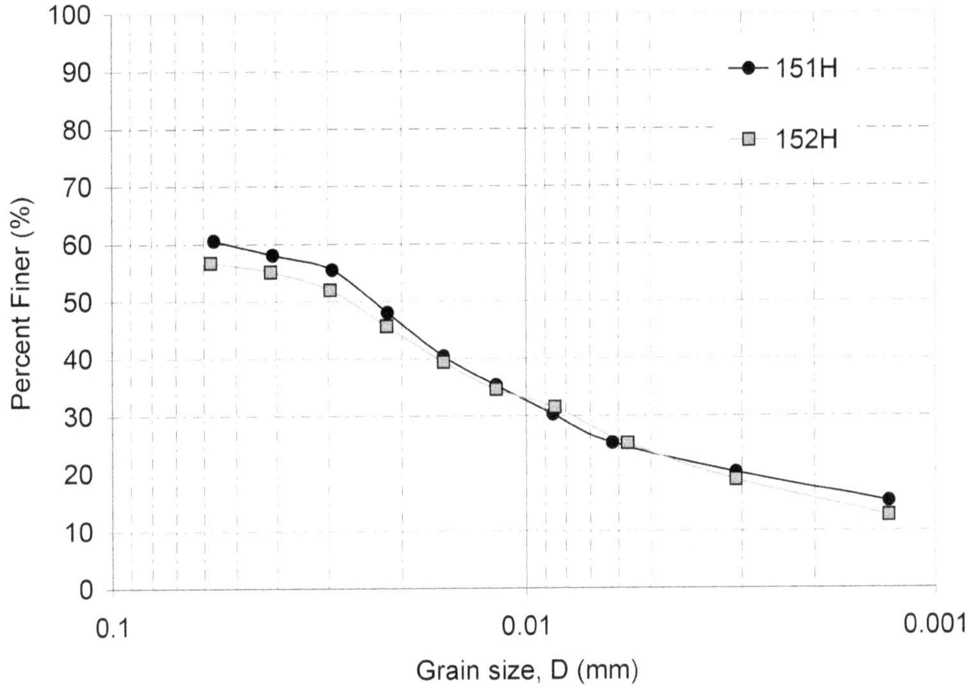


Figure 3-4. Hydrometer test results on a sandy clay using 151H and 152H hydrometers.

The figure shows that the 151H and 152H hydrometers obtained similar particle size distributions. The highest percentage is 60% due to the sand portion in the sandy clay that settles within the first 0.5 minutes and, therefore, cannot be measured. A sieve analysis should be conducted to obtain the particle size distribution, which is combined with the particle size distribution from the hydrometer test.

Name ______________________________

# Test 3: Particle Size Distribution of Soils – Hydrometer Test

Question 1: What are the common types of hydrometers used in a hydrometer test?

Question 2: What dispersing agent is used in a hydrometer test? What is its concentration?

Question 3: What are the typical soil sizes and masses used in a hydrometer test?

Question 4: What are the three corrections for a hydrometer reading?

# TEST 4

# Specific Gravity of Soil Solids

## 1. Introduction

The objective of this test is to determine the specific gravity of soil solids. Specific gravity, $G_s$, is defined as

$$G_s = \frac{\rho_s}{\rho_w}$$

where:

$\rho_s$ = density of soil solids, and

$\rho_w$ = density of distilled water at 4°C (1 g/cm$^3$).

Based on the definition, $G_s$ can be defined as the mass of the soil solids divided by the mass of the water with an equal volume:

$$G_s = \frac{\rho_s}{\rho_w} = \frac{\rho_s V}{\rho_w V} = \frac{M_{solid}}{M_{water\_of\_same\_volume\_as\_solid}}$$

The mass of the solids is determined by oven drying, and the volume of the solid is determined by the volume of water that the solids displace. To accurately measure the volume of water displaced by the solids, any air introduced by the soil when submerged in water should be removed. Therefore, the test involves de-airing the soil-water mixture using boiling or vacuuming.

## 2. References


- ASTM D 854-05 "Standard Test Methods for Specific Gravity of Soil Solids by Water Pycnometer," *2006 Annual Book of ASTM Standards*, Volume 04.08, Soil and Rock (I). ASTM International.
- ASTM D 5550-00 "Standard Test Methods for Specific Gravity of Soil Solids by Gas Pycnometer," *2006 Annual Book of ASTM Standards*, Volume 04.08, Soil and Rock (I). ASTM International.

## 3. Soils and Other Materials

- Any type of soil (air dry)
- Distilled water (about 1 L)
- De-aired water (about 1 L)

## 4. Apparatus

- One pycnometer (500 mL volumetric flask with a 500 mL mark)
- Mixing container (beaker or porcelain bowl with a spatula or straightedge)
- A large pan or bowl
- Balance of 0.01 g readability
- Drying oven maintaining a constant temperature of 110±5°C
- Thermometer of 0.1°C readability
- De-airing apparatus (burner or vacuum system)
- Funnel with a stem that extends past the calibration mark on the volumetric flask
- No. 4 (4.75 mm) sieve
- Wash bottle
- Insulated container to keep constant temperature (not needed if the test is conducted in a constant temperature room)

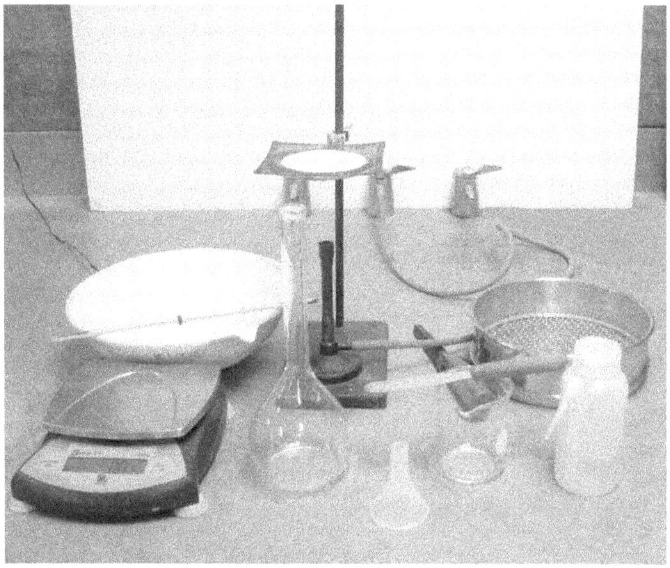

Figure 4-1. Specific gravity measurement apparatus.

## 5. Preparation of Test Specimen

- Moist (or air-dried) soil that passes the No. 4 sieve.
- Recommended **dry** mass of test specimen: sandy soil (can contain clay or silt): 100±10 g; silt or clay: 50±10 g

## 6. Test Procedure

(1) Measure the mass of a dry and clean volumetric flask.
(2) Add de-aired water to the dry and clean flask to the 500 mL mark. Measure and record the mass; then empty the flask.

(3) Measure the recommended amount of soil and 100 mL of distilled water. Mix the soil and the water well in a mixing container.
(4) Using a funnel, pour the slurry into the volumetric flask (Figure 4-2). Using a wash bottle, rinse any soil particles remaining in the mixing container and on the funnel into the flask.
(5) Add water into the flask until the water level is between 1/3 or 1/2 of the depth of the main body of the flask. Agitate the water until a slurry is formed. Rinse any soil adhering to the flask into the slurry.
(6) De-airing the soil slurry:
   - **Heating (boiling)** (Figure 4-3). Heat the flask with the slurry using a burner for at least 2 hours after the soil-water mixture comes to a full boil. Use only enough heat to keep the slurry boiling. Agitate the flask periodically to prevent any soil from sticking to or drying onto the glass.

Figure 4-2. Pouring the slurry into the flask.

Figure 4-3. De-airing the slurry using boiling.

Figure 4-4. Cooling.

Figure 4-5. Slurry in a bowl before oven-drying.

   - **Vacuum**: The vacuum must remain relatively constant and be sufficient to cause bubbling at the beginning of the de-airing process. Agitate the flask continually under the vacuum for at least 2 hours.
   - **Combination of Heating and Vacuum**: Follow the above approaches with a duration of at least 1 hour.

(7) Fill the flask with de-aired water.
  - Use a small-diameter flexible tubing with one end just below the surface of the slurry and add water through the tubing to prevent air from being introduced into the slurry.
  - Do not add water above the calibration mark. If the water in the flask becomes cloudy, stop. Add the remaining water when the water in the neck becomes clear.

(8) If using the boiling method, cover the flask and let it cool to room temperature (this may take from several hours to overnight) (Figure 4-4).
After cooling, adjust the water to the calibration mark. Only touch the rim of the flask.

(9) Dry the rim using a paper towel.

(10) Measure and record the mass of the flask and the soil-water mixture.

(11) Measure the temperature of the soil-water mixture using a thermometer.

(12) Transfer the slurry into a large pan or bowl. Wash all the soil solids in the flask into the pan or bowl (Figure 4-5).

(13) Dry the slurry to a constant mass in an oven at 110±5°C overnight or until a constant mass is reached.

(14) Measure and calculate the mass of the dry soil.

## 7. Data Collection and Calculation

Mass of dry and clean flask, $M_{\text{flask}}$ =

Mass of flask with de-aired water to the 500 mL mark, $M_{\text{flask+water}}$ =

Mass of flask with soil-water mixture to the 500 mL mark, $M_{\text{flask+water+soil}}$ =

Mass of large bowl (dry), $M_1$ =

Mass of large bowl with soil after oven drying, $M_2$ =

Mass of oven-dried soil after the test, $M_{\text{soil\_solid}} = M_2 - M_1$ =

Temperature, $T$ =

## 8. Results and Analyses

Calculate the specific gravity at temperature $T$:

$$G_s = \frac{M_{soil\_solid}}{M_{water\_of\_same\_volume\_as\_solid}} = \frac{M_{soil\_solid}}{M_{flask+water} - (M_{flask+water+soil} - M_{soil\_solid})}$$

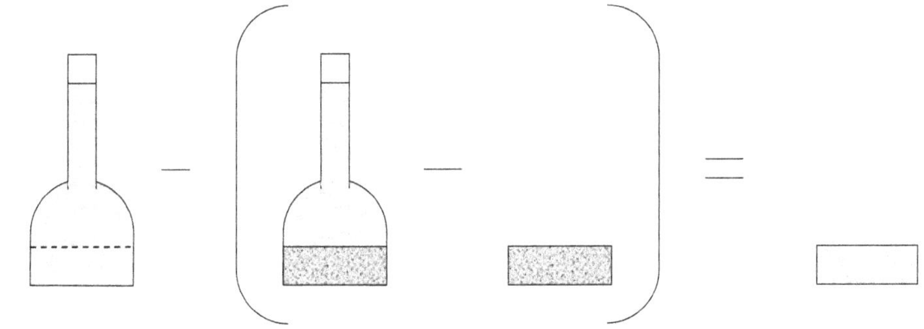

$$M_{flask+water} - \left[ M_{flask+water+soil} - M_{soil\_solid} \right] = M_{water\_of\_same\_volume\_as\_solid}$$

Figure 6. Calculation of mass of water with the same volume as soil solids.

If required, calculate the specific gravity of soil solids at 20°C:

$$G_{s(20^{o}C)} = \frac{\rho_{w(T^{o}C)}}{\rho_{w(20^{o}C)}} \cdot G_{s(T)}$$

where:

$\rho_{w(T^{o}C)}$ = water density at T°C, and

$\rho_{w(20^{o}C)}$ = water density at 20°C. The values can be found in Table 4-1.

Table 4-1. Water Density at Different Temperatures

| **Temperature (°C)** | **Density (g/cm³)** | **Temperature (°C)** | **Density (g/cm³)** |
|---|---|---|---|
| 15 | 0.999099 | 23 | 0.997538 |
| 16 | 0.998943 | 24 | 0.997296 |
| 17 | 0.998774 | 25 | 0.997044 |
| 18 | 0.998595 | 26 | 0.996783 |
| 19 | 0.998405 | 27 | 0.996512 |
| 20 | 0.998203 | 28 | 0.996232 |
| 21 | 0.997992 | 29 | 0.995944 |
| 22 | 0.997770 | 30 | 0.995646 |

## SAMPLE DATA COLLECTION, RESULTS, AND ANALYSES

Mass of dry and clean flask, $M_{\text{flask}}$ = 181.10 g
Mass of flask with de-aired water to the 500 mL mark, $M_{\text{flask+water}}$ = 681.03 g
Mass of flask with soil-water mixture to the 500 mL mark, $M_{\text{flask+water+soil}}$ = 710.82 g
Mass of large bowl (dry), $M_1$ = 888.84 g
Mass of large bowl with soil after oven drying, $M_2$ = 936.63 g
Mass of oven-dried soil after the test, $M_{\text{soil\_solid}} = M_2 - M_1$ = 47.79 g
Temperature, $T$ = 22.8°C

$$G_s = \frac{M_{soil\_solid}}{M_{water\_of\_same\_volume\_as\_solid}} = \frac{M_{soil\_solid}}{M_{flask+water} - (M_{flask+water+soil} - M_{soil\_solid})}$$

$$= \frac{47.79}{681.03 - (710.82 - 47.79)} = 2.66$$

Name ______________________________

# Test 4: Specific Gravity of Soil Solids

Question 1: What is the definition of specific gravity?

Question 2: What methods can be used to de-air the water?

Question 3: Why is it important to remove air from the water?

# TEST 5

# Compaction Test of Soils

## 1. Introduction

The objective of this test is to determine the optimum water content and the maximum dry unit weight (or density) of soils.

Engineered fills, such as embankments, foundations, and road bases, are compacted to a dense state to obtain satisfactory shear strength, compressibility, or permeability. Adding water to dry soil facilitates the compaction, as water acts as a lubricant for soil particle rearrangement. Too much water in the voids, however, may reduce the density under the same compaction energy, as pore water pressure resists compaction and water does not easily vacate the pore. Therefore, the density of soils, represented by dry density, varies with the water content under the same compaction energy. The relationship is shown in Figure 5-1.

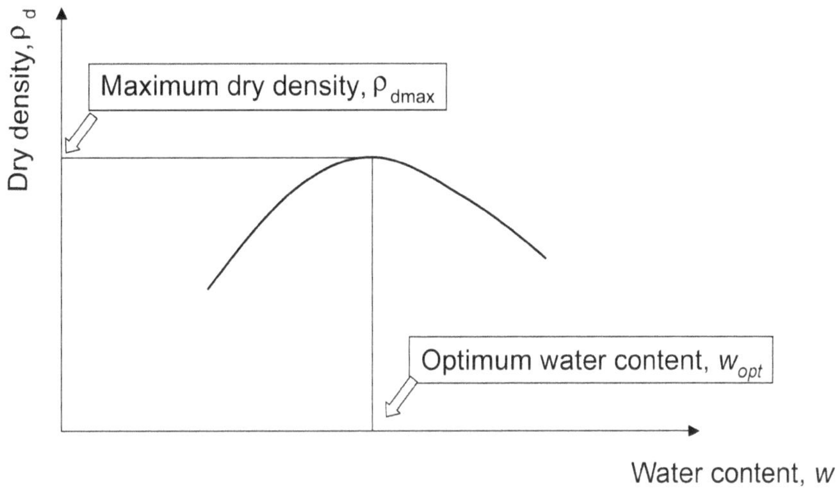


Figure 5-1. Compaction curve.

The peak value of the dry density is the maximum dry density for a given compaction energy, $\rho_{dmax}$, which occurs at a water content called the optimum water content, $w_{opt}$.

In this test, soil specimens at different water contents are compacted using the same compaction effort and dry density after compaction is determined. The compaction curve can then be plotted to obtain $w_{opt}$ and $\rho_{dmax}$.

## 2. References

- ASTM D 698-00 "Standard Test Methods for Laboratory Compaction Characteristics of Soil Using Standard Effort (12,400 ft-lbf/ft$^3$ (600 kN-m/m$^3$))," *2006 Annual Book of ASTM Standards*, Volume 04.08, Soil and Rock (I). ASTM International.
- ASTM D 1557-02 "Standard Test Methods for Laboratory Compaction Characteristics of Soil Using Modified Effort (56,000 ft-lbf/ft$^3$ (2,700 kN-m/m$^3$))," *2006 Annual Book of ASTM Standards*, Volume 04.08, Soil and Rock (I). ASTM International.
- ASTM D 2216-05 "Standard Test Methods for Laboratory Determination of Water (Moisture) Content of Soil and Rock by Mass," *2006 Annual Book of ASTM Standards*, Volume 04.08, Soil and Rock (I). ASTM International.
- ASTM D 4643-00 "Standard Test Methods for Laboratory Determination of Water (Moisture) Content of Soil by the Microwave Oven Heating," *2006 Annual Book of ASTM Standards*, Volume 04.08, Soil and Rock (I). ASTM International.
- ASTM D 4959-00 "Standard Test Methods for Laboratory Determination of Water (Moisture) Content of Soil by Direct Heating," *2006 Annual Book of ASTM Standards*, Volume 04.08, Soil and Rock (I). ASTM International.

## 3. Soils and Other Materials

- Any type of soil

Figure 5-2. Compaction apparatus.

## 4. Apparatus

- Mold assembly, including mold, base plate, and extension collar. The mold used in this test is 4 inches in diameter, although a 6-inch diameter mold can also be used per ASTM D 698-00 and ASTM D 1557-02 for soils that have more than 20% retained on the 3/8-inch sieve.
- Rammer: 5.5-lbf for *standard* test; 10.0-lbf for *modified* test
- Sample extruder: for clay soil, a sample extruder may be used to extrude the compacted hard sample; for sandy soil, the samples can be easily extruded using a screwdriver
- Water content containers (such as porcelain dishes) to measure water content
- Balances: one with a readability of 0.01 g for water content measurement; one with a readability of 1 g and a capacity of 10 kg to measure the mass of the mold with the soil

- Mixing pans or bowls
- Spatula for soil mixing
- Sieve (8-inch diameter, No. 4 or 4.75 mm)
- Straightedge
- Wash bottle
- Drying oven

## 5. Preparation of Test Specimen

Note:

- Moist preparation method is used, without previously drying the sample.
- ASTM specifies that soil that has been previously compacted in the laboratory should not be re-used. If this test is performed in a teaching laboratory where a sufficient amount of soil is not available, compacted soils can be reused for demonstration and teaching purpose only.

(1) Collect 25 lbm (11.4 kg) of soils that pass the No. 4 sieve. Divide the soil evenly and place it in five mixing pans or bowls (5 lbm or 2.2 kg per bowl). Record the wet soil mass in each bowl, $M_{wet}$.

(2) Measure the initial water content of the soil, $w_{ini}$. If the water content is higher than 5%, the soil should be air-dried and the initial water content re-measured. For rapid measurement, direct heating (ASTM D4959) can be used.

(3) Specify the target water content for each of the five specimens, to provide at least two specimens wet and two specimens dry of optimum water content. The water content increment should be about 2% and should not exceed 4%.

(4) Calculate the dry soil mass in each of the mixing bowls, using

$$M_{dry} = \frac{M_{wet}}{1 + \frac{w_{ini}(\%)}{100\%}}$$

(5) Calculate the amount of water ($\Delta M_w$) to add to each specimen to reach the water content specified in (3), using

$$\Delta M_w = (w_{specify} - w_{ini}) \times M_{dry}$$

(6) Thoroughly mix the soil in each container, and cover it with lid (or aluminum foil or wet towel) and allow it to stand (cure) for 3 hours if the soil is gravel and sand and 16 hours if the soil is silt or clay.

## 6. Test Procedure

**Compaction Using Standard Effort (Standard Proctor Test):**

(1) After the curing, remix the soil thoroughly in each mixing bowl.

(2) Secure the mold to the base plate and measure the mass of the mold and the base plate.

(3) Attach and secure the collar extension to the mold.

Figure 5-3. Compaction.

Figure 5-4. Specimen after compaction.

(4) Compact the specimen in **three** layers using the standard rammer (5.5 lbf), **25** blows per layer.
- For each layer, place approximately the same amount of soil so that each layer is approximately equal in thickness after compaction.
- Prior to compaction, place the loose soil into the mold and spread it evenly. Lightly tamp the soil so it is not fluffy.
- Hold the guide sleeve steady and vertically lift the inner rammer all the way up and let it free-fall on the soil. Move the rammer around in the mold to provide uniform compaction.
- Apply the blows at a uniform rate of approximately 25 blows/minute.
- After the compaction of each layer, any soil that is adjacent to the mold walls and has not been compacted or extends above the compacted surface shall be trimmed using a knife or straightedge. The trimmed soil may be included with the additional soils for the next layer.
- The third layer should slightly extend into the collar by about 5 mm, but does not exceed 1/4 inch (6 mm) above the top of the mold. If the third layer does extend above the top of the mold by 1/4 inch or is below the top of the mold, the specimen should be discarded. The soil should be returned to the mixing pan and thoroughly mixed; then start another compaction.

(5) Carefully remove the collar from the mold and avoid disrupting the soil below the top of the mold. Trim the specimen to be even with the top of the mold using a straightedge, taking care not to tear the soil below the top of the mold.

(6) Measure the mass of the mold and the base plate with the soil.

(7) Extrude the soil from the mold, using a sample extruder.

(8) Obtain a sample for water content measurement. Put the water content container in the oven at 110°C. Two sampling methods can be used:
- Use the entire specimen (preferred method); break up the specimen to facilitate drying.
- Slice the specimen axially through the center and take 500 g of sample from the cut faces. Porcelain dishes can be used as containers.

(9) Repeat the above steps for the other four specimens.

(10) Clean the apparatus. Determine the water content the next day. Discard the soil in the containers.

**Compaction Using Modified Effort (Modified Proctor Test):**

There are two differences between the Standard Proctor Test and the Modified Proctor Test:

1. The rammer in the Standard Proctor Test is 5.5 lbf and has a 1.0-ft drop; in the Modified Proctor Test, it is 10.0 lbf and has a 1.5-ft drop.
2. The number of layers in the Standard Proctor Test is 3; in the Modified Proctor Test, the number of layers is 5.

The rest of the test procedures are the same as in the Standard Proctor Test.

3. Data Collection and Calculation

Table 5-1. Initial Water Content Measurement

| | |
|---|---|
| Mass of moisture container, $M_1$ (g) | |
| Mass of container + moist soil, $M_2$ (g) | |
| Mass of container + dry soil, $M_3$ (g) | |
| Water content, $w(\%) = \frac{M_2 - M_3}{M_3 - M_1} \times 100\%$ | |

Table 5-2. Compaction Tests

| | 1 | 2 | 3 | 4 | 5 |
|---|---|---|---|---|---|
| Mass of mold + base plate, $M_1$ (g) | | | | | |
| Mass of mold, base plate, and moist soil, $M_2$ (g) | | | | | |
| Mass of moist soil, $M_2 - M_1$ (g) | | | | | |
| Mass of moist soil, $M_2 - M_1$ (lbm) (1 g = 0.0022 lbm) | | | | | |
| Moist unit weight (the volume of the mold is 1/30 ft$^3$), $\gamma = \frac{M_2 - M_1}{1/30}$ (lb/ft$^3$) | | | | | |
| Mass of moisture container, $M_3$ (g) | | | | | |
| Mass of container + moist soil, $M_4$ (g) | | | | | |
| Mass of container + dry soil, $M_5$ (g) | | | | | |
| Water content, $w(\%) = \frac{M_4 - M_5}{M_5 - M_3} \times 100\%$ | | | | | |
| Dry unit weight, $\gamma_d(\text{lb/ft}^3) = \frac{\gamma}{1 + \frac{w(\%)}{100\%}}$ | | | | | |

## 4. Results and Analyses

(1) Complete the calculation in Tables 5-1 and 5-2.
(2) Plot the compaction curve of the dry unit weight and the water content, using the data in Table 5-2.
(3) From the curve, identify and report the maximum dry unit weight and the optimum water content.
(4) Plot the 100% saturation curve (a.k.a., zero air void curve, i.e., all the air is expelled) along with the compaction curve. If the specific gravity ($G_s$) is unknown, assume a value based on the soil type. The dry unit weight of a specimen at 100% saturation after compaction can be calculated using

$$\gamma_{zav} = \frac{\gamma_w}{\frac{\omega(\%)}{100\%} + \frac{1}{G_s}}$$

**Table 5-3. Zero Air Void Curve** ($\gamma_w$= 62.4 lb/ft$^3$)

| | | | | | |
|---|---|---|---|---|---|
| Water content, $w$ (%) | | | | | |
| $\gamma_{zav}$ (lb/ft$^3$) | | | | | |

## SAMPLE DATA COLLECTION, CALCULATION, RESULTS, AND ANALYSES

**Standard Proctor Test on Sandy Clay:**

Table 5-1. Initial Water Content Measurement

| | |
|---|---|
| Mass of moisture container, $M_1$ (g) | 21.7 |
| Mass of container + moist soil, $M_2$ (g) | 127.8 |
| Mass of container + dry soil, $M_3$ (g) | 124.7 |
| Water content, $w(\%) = \frac{M_2 - M_3}{M_3 - M_1} \times 100\%$ | 3.0 |

Table 5-2. Compaction Tests

| | 1 | 2 | 3 | 4 | 5 |
|---|---|---|---|---|---|
| Mass of mold + base plate, $M_1$ (g) | 4255 | 4255 | 4255 | 4255 | 4255 |
| Mass of mold, base plate, and moist soil, $M_2$ (g) | 5917 | 5977 | 6067 | 6118 | 6068 |
| Mass of moist soil, $M_2 - M_1$ (g) | 1662 | 1722 | 1812 | 1863 | 1813 |
| Mass of moist soil, $M_2 - M_1$ (lbm) (1 g = 0.0022 lbm) | 3.6564 | 3.7884 | 3.9864 | 4.0986 | 3.9886 |
| Moist unit weight (the volume of the mold is 1/30 ft³), $\gamma = \frac{M_2 - M_1}{1/30}$ (lb/ft³) | 109.7 | 113.6 | 119.6 | 123.0 | 119.7 |
| Mass of moisture container, $M_3$ (g) | 157.5 | 152.5 | 328.5 | 157.2 | 157.5 |
| Mass of container + moist soil, $M_4$ (g) | 364.7 | 344.5 | 551 | 395.1 | 438.3 |
| Mass of container + dry soil, $M_5$ (g) | 338.8 | 318.9 | 518.1 | 356.1 | 388.2 |
| Water content, $w(\%) = \frac{M_4 - M_5}{M_5 - M_3} \times 100\%$ | 14.3 | 15.4 | 17.4 | 19.6 | 21.7 |
| Dry unit weight, $\gamma_d(\text{lb/ft}^3) = \frac{\gamma}{1 + \frac{w(\%)}{100\%}}$ | 96.0 | 98.5 | 101.9 | 102.8 | 98.3 |

**Modified Proctor Test on the Same Sandy Clay:**

Table 5-1. Initial Water Content Measurement

| | |
|---|---|
| Mass of moisture container, $M_1$ (g) | 21.7 |
| Mass of container + moist soil, $M_2$ (g) | 127.8 |
| Mass of container + dry soil, $M_3$ (g) | 124.7 |
| Water content, $w(\%) = \dfrac{M_2 - M_3}{M_3 - M_1} \times 100\%$ | 3.0 |

Table 5-2. Compaction Tests

| | 1 | 2 | 3 | 4 | 5 |
|---|---|---|---|---|---|
| Mass of mold + base plate, $M_1$ (g) | 2041 | 2041 | 2041 | 2041 | 2041 |
| Mass of mold, base plate, and moist soil, $M_2$ (g) | 3828 | 3891 | 3947 | 3983 | 3948 |
| Mass of moist soil, $M_2 - M_1$ (g) | 1787 | 1850 | 1906 | 1942 | 1907 |
| Mass of moist soil, $M_2 - M_1$ (lbm) (1 g = 0.0022 lbm) | 3.9322 | 4.0698 | 4.1932 | 4.2713 | 4.1958 |
| Moist unit weight (the volume of the mold is 1/30 ft$^3$), $\gamma = \dfrac{M_2 - M_1}{1/30}$ (lb/ft$^3$) | 118.0 | 122.1 | 125.8 | 128.1 | 125.9 |
| Mass of moisture container, $M_3$ (g) | 27.9 | 27.5 | 21.9 | 21.9 | 22.2 |
| Mass of container + moist soil, $M_4$ (g) | 176.9 | 84.2 | 87.4 | 121.6 | 116.5 |
| Mass of container + dry soil, $M_5$ (g) | 163.6 | 78.2 | 79.4 | 108.1 | 102.3 |
| Water content, $w(\%) = \dfrac{M_4 - M_5}{M_5 - M_3} \times 100\%$ | 9.8 | 11.8 | 13.9 | 15.7 | 17.7 |
| Dry unit weight, $\gamma_d(\text{lb/ft}^3) = \dfrac{\gamma}{1 + \dfrac{w(\%)}{100\%}}$ | 107.4 | 109.2 | 110.6 | 110.8 | 107.0 |

Table 5-3. Zero Air Void Curve

$G_s$ = 2.65

$\gamma_w$ = 62.4 lb/ft$^3$

| | | | | | |
|---|---|---|---|---|---|
| Water content, $w$ (%) | 15 | 17 | 19 | 21 | 23 |
| $\gamma_{zav}$ (lb/ft$^3$) | 118.3 | 114.0 | 110.0 | 106.2 | 102.7 |

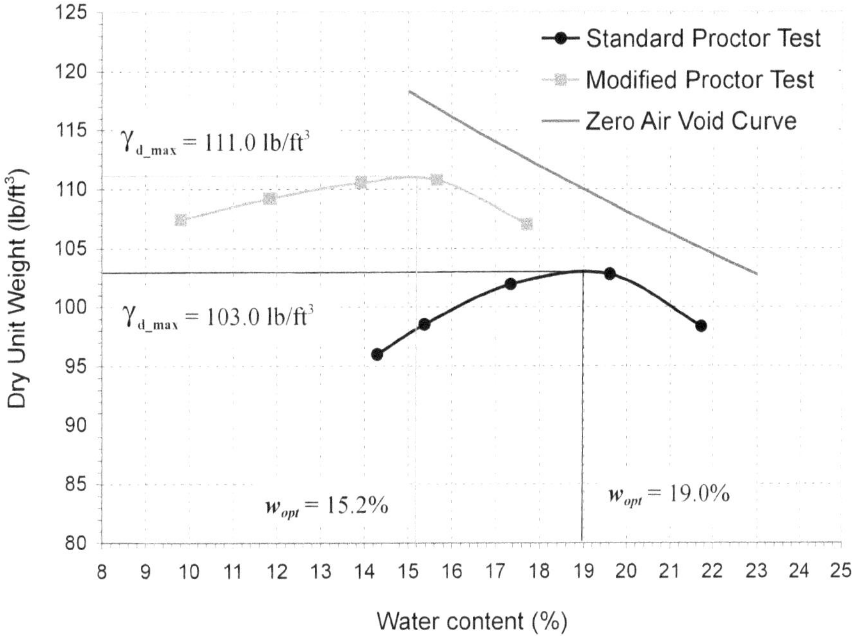


Figure 5-5. Compaction test results on sandy clay.

Name ______________________________

# Test 5: Compaction Test of Soils

Question 1: What is the purpose of a compaction test?

Question 2: What are the differences between the standard Proctor test and the modified Proctor test?

Question 3: How should a soil be cured before the compaction test?

Question 4: In a compaction test, the moist unit weight of a soil specimen is determined to be 19 kN/m$^3$, the moisture content of the specimen is 12%. What is the dry unit weight of the specimen?

T E S T 6

# Liquid Limit and Plastic Limit of Soils

## 1. Introduction

The objective of this test is to determine the liquid limit and the plastic limit of cohesive soils. With different water content, soils may manifest four consistency states: solid, semi-solid, plastic, and liquid (Figure 6-1). The boundary between the solid and the semi-solid states, in terms of water content, is the shrinkage limit (SL); the boundary between the semi-solid and the plastic states is the plastic limit (PL); and the boundary between the plastic and the liquid states is the liquid limit (LL). The SL, PL, and LL are collectively referred to as the Atterberg Limits in recognition of their formation by a Swedish soil scientist, Albert Atterberg.

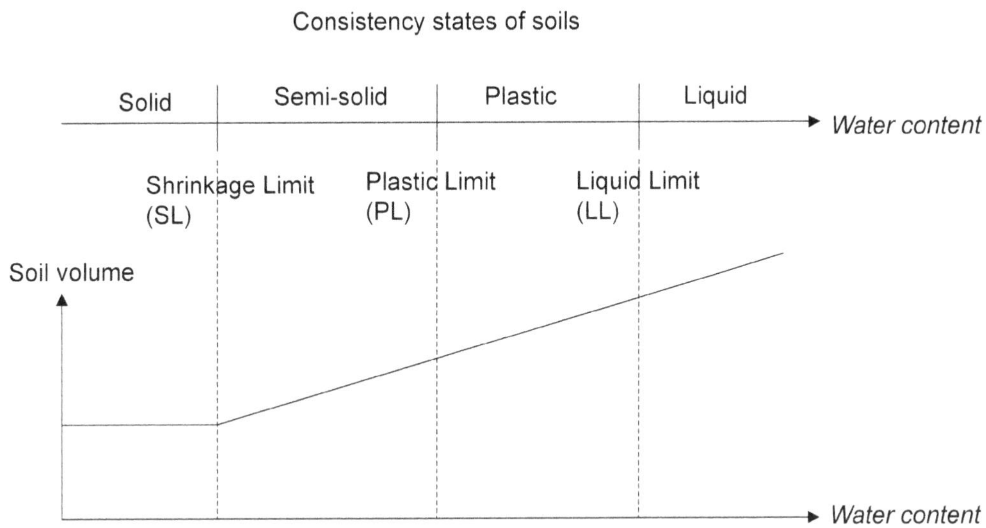


Figure 6-1. Consistency states of soils due to different water content.

## 2. References

- ASTM D 4318-05 "Standard Test Methods for Liquid Limit, Plastic Limit, and Plasticity Index of Soils," *2006 Annual Book of ASTM Standards*, Volume 04.08, Soil and Rock (I). ASTM International.

## 3. Soils and Other Materials

- Clayey soil
- Distilled or de-mineralized water per ASTM D4318-05

## 4. Apparatus

- Liquid limit device, including a hard rubber base, rubber feet to support the base, brass cup, and cam to raise the cup smoothly and continuously. The liquid limit device can be motorized or manual.

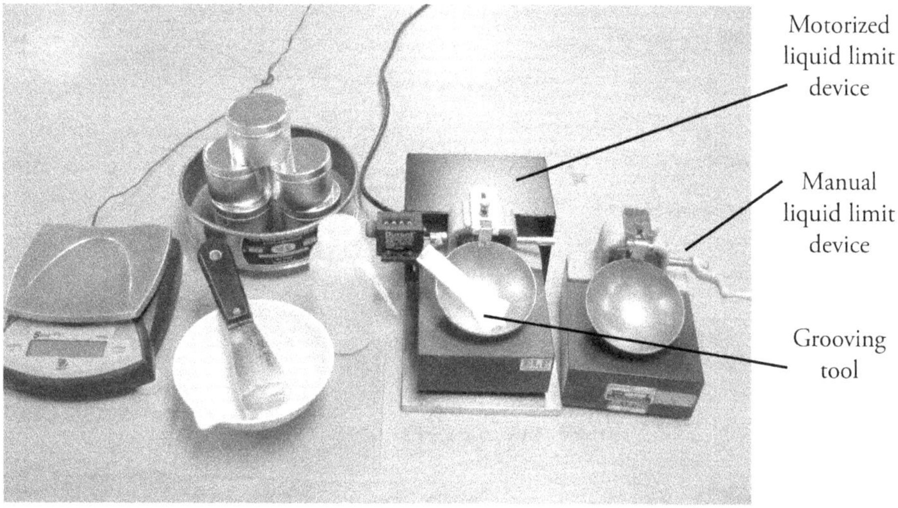


Figure 6-2. Apparatus for liquid and plastic limits tests.

- Flat grooving tool to make a groove in the soil specimen
- Water content containers (moisture can) to measure water content
- Balance with a readability of 0.01 g
- Mixing dish
- Ground-glass plate
- Spatula for soil mixing
- Sieve (8-inch diameter, No. 40 or 425 μm)
- Wash bottle
- Drying oven

## 5. Preparation of Test Specimen

(1) Collect 150 to 200 g of air-dried soil that passes the No. 40 (425 μm) sieve.

(2) Mix the soil thoroughly with water in a mixing dish. Adjust the water content of the soil to a consistency that would require about 25 to 35 blows of the liquid limit device to close the groove by 13 mm (see the following procedure for operation instructions). Initial mixing time of more than 30 minutes may be needed for stiff, fat clays.

(3) Place the prepared soil in the mixing dish, cover it to prevent moisture loss, and allow the soil to stand (cure) for at least 16 hours (overnight).

(4) At the end of the curing, set aside about 20 g of soil for plastic limit measurement.

## 6. Test Procedure

This test follows Method A, Multipoint Test, using the wet preparation method, specified by ASTM D 4318-05.

Figure 6-3. Soil pat.

Figure 6-4. With groove.

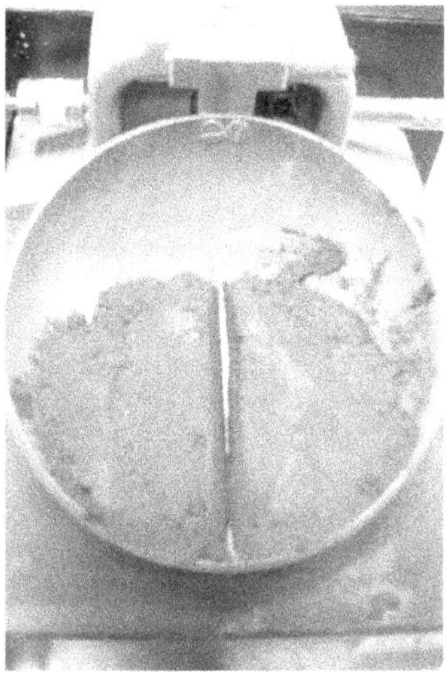

Figure 6-5. 0.5-inch closure.

**Liquid Limit Measurement:**

(1) Adjust the height-of-drop of the cup to 10±0.2 mm. The grooving tool can be used, as most grooving tools have a tail that is 10 mm.

(2) Thoroughly remix the soil before the test. If needed, re-adjust its water content until the consistency requires about 25 to 35 blows of the liquid limit device to close the groove by 13 mm (see the following procedure for operation instructions).

(3) Using a spatula, place a portion of the prepared soil in the brass cup of the liquid limit device at the point where the cup rests on the base; squeeze the soil down and spread it into the cup to a depth of about 10 mm at its deepest point. The soil should cover about 2/3 of the cup area at the lower side of the cup. Taper the soil to form a flat surface (Figure 6-3).

(4) Keep the unused soil in the mixing dish and cover it with a wet towel to retain the moisture while performing the test.

(5) Form a groove in the middle of the soil pat using the grooving tool. When cutting the groove, hold the grooving tool against the cup surface and draw an arc, maintaining the tool perpendicular to the cup surface (Figure 6-4).
   - The soil depth is such that the grooving tool should barely trim the soil pat at the deepest part of the cup.
   - If the shoulders of the grooving tool trim more than 2 mm of soil at the deepest point of the soil pat, there is too much soil; if the shoulders of the grooving tool do not touch the soil pat at any point, there is not enough soil. In either scenario, do not simply take some soil off or add some soil; remove the soil, clean the cup, and start over.
   - Take caution to prevent sliding of the soil pat against the cup surface while making the groove.

(6) Raise and drop the cup to the base.
   - If using a manual liquid limit device, rotate the crank at a rate of 2 drops per second until the two halves of the soil pat come in contact at the bottom of the groove with a length of 13 mm (0.5 inches). Count the number of drops, *N*.

- If using a motorized liquid limit device with a counter, switch it on. Switch it off as soon as the groove closure is 13 mm (0.5 inches). The counter records the number of drops (Figure 6-5).
- The first test should have an $N$ value of **25 to 35.**

(7) Measure the water content of the soil pat.
- Measure the mass of a moisture can.
- Collect a slice of soil that is perpendicular to the groove and passes the soil closure; put the soil into the moisture can and measure the mass.
- Put the can into the oven.

(8) Return the remainder of the soil pat to the mixing dish. Wash and dry the cup and the grooving tool and re-attach the cup to the liquid limit device. Add a small amount of water and remix the soil thoroughly. Repeat Steps (3) through (7). This time, the number of drops should be **20 to 30**.

(9) Repeat the test again, following the instructions of Steps (3) through (7). This time, the number of drops should be **15 to 25**.

(10) Clean and put away the liquid limit device. Come back the second day to measure the masses of the moisture cans in the oven. Discard the soils in the cans and clean the cans.

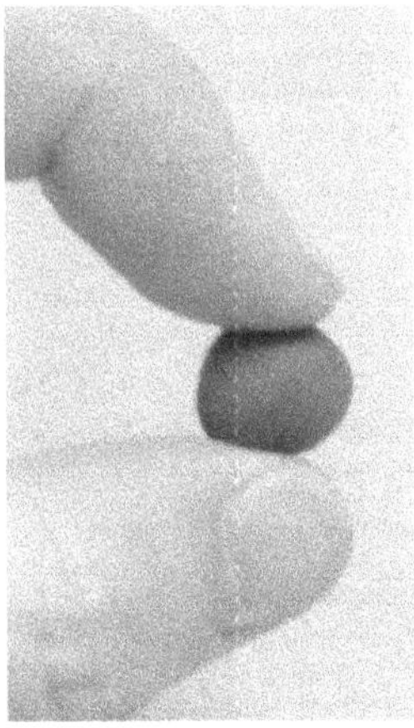
Figure 6-6. Soil ellipsoid.

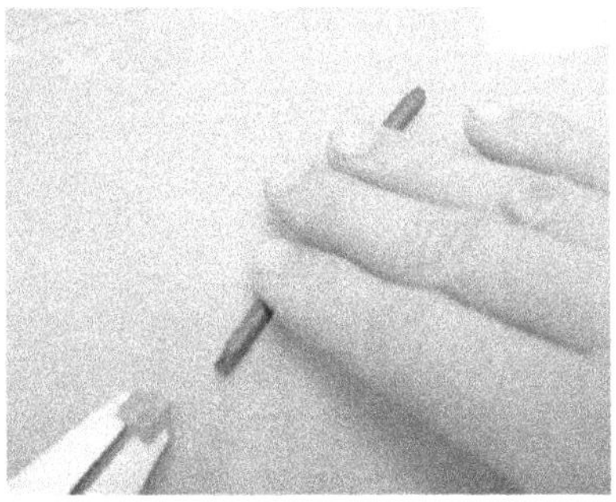
Figure 6-7. Rolling the thread.

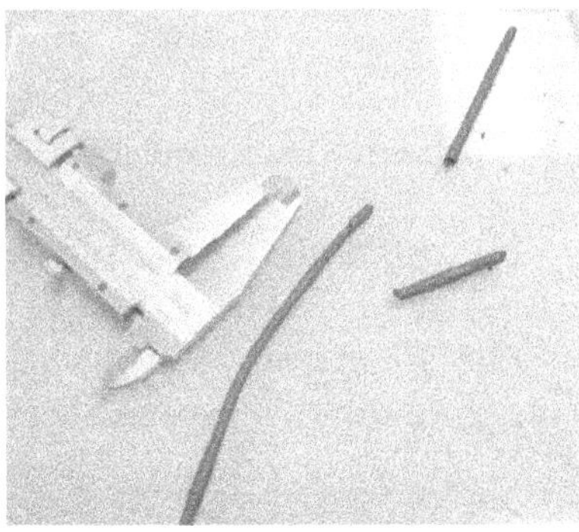
Figure 6-8. Thread breakup.

**Plastic Limit Measurement:**

(1) From the 20 g of the soil reserved for the plastic limit measurement, take about 2 g of soil, and form it into an ellipsoidal shape (Figure 6-6).

(2) Roll the soil mass into a thread with uniform diameter of **3.2 mm (1/8 inch)**, taking less than 2 minutes (Figure 6-7).
- Two methods can be used to roll the soil into a thread: the hand method and the rolling device method. In the hand method, roll the soil on a ground-glass plate using the palm or finger. The rolling device method is specified per ASTM D 4318-05.
- If the soil thread does not break (crumble) at 3.2 mm, break the soil thread into several pieces, gather the crumbles, and re-roll the soil into a 3.2 mm thread. Repeat this step until the soil thread crumbles at exactly 3.2 mm.
- If the soil thread crumbles before it reaches 3.2 mm diameter, return the soil thread to the remaining soil, add a slight amount of water, and remix it thoroughly. Then repeat Steps (1) and (2) until the soil thread crumbles at exactly 3.2 mm.

(3) Collect the soil pieces, immediately transfer them into a moisture can, and close the lid.

(4) Repeat Steps (1) and (2) twice to collect about **6 g** of soil in the moisture can.

(5) Repeat Steps (1) through (4) to collect 6 g of soils in a *second* moisture can.

(6) Measure the masses of the two moisture cans with soil. Put the two moisture cans in the oven. Clean and put away the ground-glass plate. Come back the second day to measure the mass of the moisture cans with the dry soil in the oven. Discard the soils in the cans and clean the cans.

## 7. Data Collection and Calculation

Table 6-1. Liquid Limit Measurement

| | **Test 1** | **Test 2** | **Test 3** |
|---|---|---|---|
| Number of drops, $N$ | | | |
| Mass of moisture can (g) | | | |
| Mass of moisture can + wet soil (g) | | | |
| Mass of moisture can + soil after oven drying (g) | | | |
| Mass of water (g) | | | |
| Mass of dry soil (g) | | | |
| Moisture content (%) | | | |

Table 6-2. Plastic Limit Measurement

| | **Test 1** | **Test 2** |
|---|---|---|
| Mass of moisture can (g) | | |
| Mass of moisture can + wet soil (g) | | |
| Mass of moisture can + soil after oven drying (g) | | |
| Mass of water (g) | | |
| Mass of dry soil (g) | | |
| Moisture content (%) | | |
| Average moisture content (%) | | |

## 8. Results and Analyses

(1) Complete the calculations in Tables 6-1 and 6-2.

(2) Plot the relationship between the water content and the number of drops in Table 6-1 on a semilogarithmic graph. The water content is the ordinate on the arithmetic scale, and the number of drops is the abscissa on the logarithmic scale. Draw a straight line that best fits the test points. The water content that corresponds to 25 drops on the line is the liquid limit. Round it to the nearest whole number.

(3) For the plastic limit measurement in Table 6-2, the average moisture content is the plastic limit. Round it to the nearest whole number. **Note:** If the moisture contents of Tests 6-1 and 6-2 are off by more than 1.4%, the plastic limit test should be performed again.

(4) Report the liquid limit (LL) and the plastic limit (PL) values. Calculate the plasticity index (PI):

$$PI = LL - PL$$

## SAMPLE DATA COLLECTION, RESULTS, AND ANALYSES

Table 6-1. Liquid Limit Measurement

| | **Test 1** | **Test 2** | **Test 3** |
|---|---|---|---|
| Number of drops, $N$ | 34 | 24 | 15 |
| Mass of moisture can (g) | 27.83 | 28.42 | 27.31 |
| Mass of moisture can + wet soil (g) | 49.65 | 53.11 | 49.53 |
| Mass of moisture can + soil after oven drying (g) | 44.36 | 46.92 | 43.83 |
| Mass of water (g) | 5.29 | 6.19 | 5.70 |
| Mass of dry soil (g) | 16.53 | 18.50 | 16.52 |
| Moisture content (%) | 32.00 | 33.45 | 34.50 |

Table 6-2. Plastic Limit Measurement

| | **Test 1** | **Test 2** |
|---|---|---|
| Mass of moisture can (g) | 38.13 | 38.52 |
| Mass of moisture can + wet soil (g) | 44.04 | 45.09 |
| Mass of moisture can + soil after oven drying (g) | 42.93 | 43.89 |
| Mass of water (g) | 1.11 | 1.20 |
| Mass of dry soil (g) | 4.80 | 5.37 |
| Moisture content (%) | 23.13 | 22.35 |
| Average moisture content (%) | 22.74 | |

(1) Complete the calculations in Tables 6-1 and 6-2.

(2) Plot the relationship between the water content and the number of drops in Table 6-1 on a semilogarithmic graph. The water content is the ordinate on the arithmetic scale, and the number of drops is the abscissa on the logarithmic scale. Draw a straight line that best fits the test points.

The water content that corresponds to 25 drops on the line is the liquid limit. Round it to the nearest whole number.

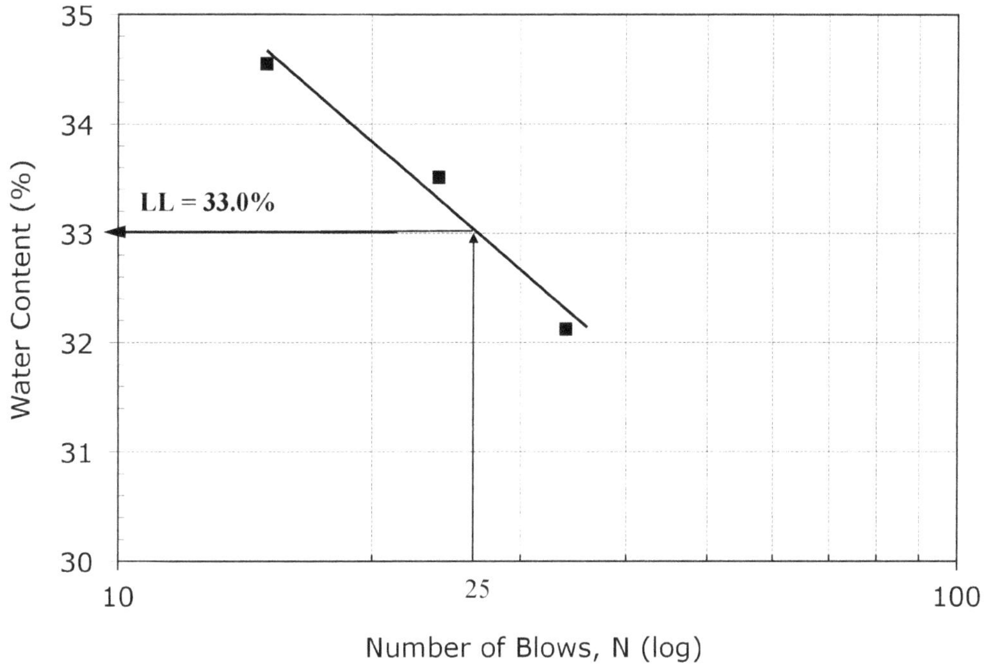


Figure 6-9. Liquid limit test results.

**Liquid Limit = 33%**

(3) For the plastic limit measurement in Table 6-2, the average moisture content is the plastic limit. Round it to the nearest whole number. **Note:** If the moisture contents of the two tests are off by more than 1.4%, the plastic limit test should be performed again.

**Plastic Limit = 23%**

(4) Report the liquid limit (LL) and the plastic limit (PL) values. Calculate the plasticity index (PI):

$$\mathbf{PI = LL - PL = 33 - 23 = 10}$$

Name ______________________

# Test 6: Liquid Limit and Plastic Limit of Soils

Question 1: What are the Atterberg Limits? What is plasticity index?

Question 2: What are the particle sizes of a soil for liquid and plastic limits measurements?

Question 3: How long should the soil be cured before the Atterberg limits measurements?

Question 4: What is the height of drop of the cup of a liquid limit device?

Question 5: What is the target diameter of a soil thread for the PL measurement?

# TEST 7

## Shrinkage Limit of Cohesive Soils

### 1. Introduction

The objective of this test is to determine the shrinkage limit of cohesive soils. The term "shrinkage limit," expressed as water content in percent, is typically assumed to represent the amount of water required to fill the voids of a given cohesive soil at its minimum void ratio obtained by drying (usually using the oven). When soil loses water, the soil volume decreases; after a particular water content is reached, further loss of water does not cause the soil volume to shrink. This water content is defined as the shrinkage limit (Figure 7-1). The concept of the shrinkage limit can be used to evaluate the shrinkage potential or possibility of development of cracks in earthworks involving cohesive soils.

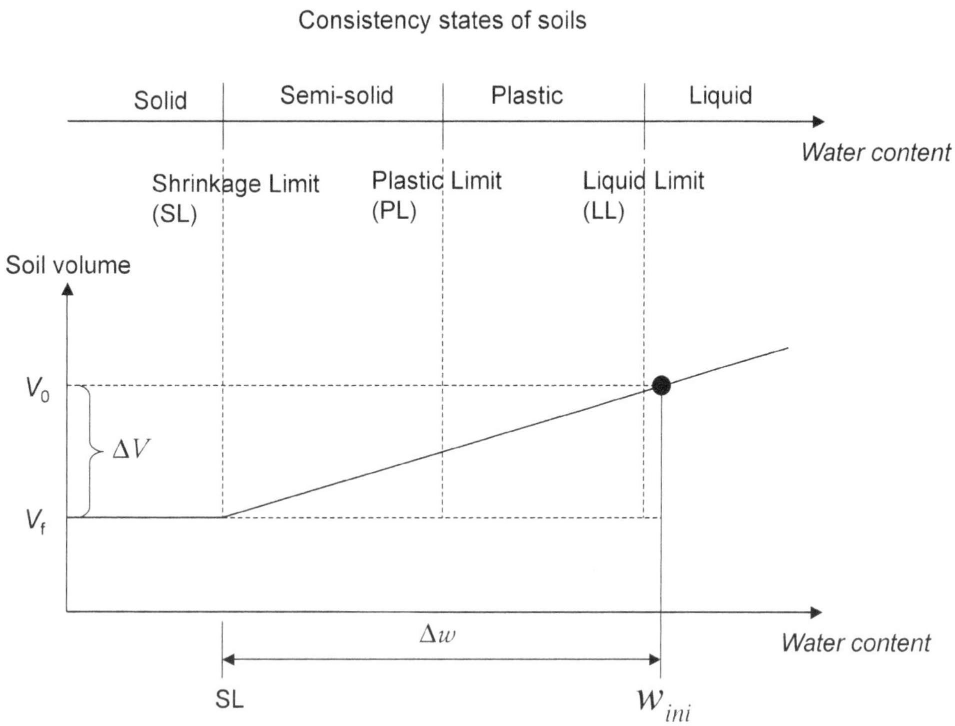


Figure 7-1. Shrinkage limit and its determination.

The measurement is based on the concept illustrated in Figure 7-1. The soil is first moistened slightly beyond the liquid limit; then a soil pat is formed in a small dish and oven-dried. The initial water content is $w_{\text{ini}}$, the portion of the water content decrement, $\Delta w$, causes the soil volume to decrease by $\Delta V$, so the shrinkage limit

$$SL = w_{ini} - \Delta w$$

where:

$$\Delta w = \rho_w \cdot \Delta V$$

$\rho_w$ = the density of water at the test temperature, and

$$\Delta V = V_0 - V_f$$

where:

$V_0$ = the initial volume of the soil pat, which equals to the volume of the dish, and
$V_f$ = the final volume of the soil pat after oven-drying (it can be measured using the wax method or the mercury method).

## 2. References

- ASTM D 4943-02 "Standard Test Methods for Shrinkage Factors of Soils by the Wax Method," *2006 Annual Book of ASTM Standards*, Volume 04.08, Soil and Rock (I). ASTM International.
- ASTM D 427-04 "Standard Test Methods for Shrinkage Factors of Soils by the Mercury Method," *2006 Annual Book of ASTM Standards*, Volume 04.08, Soil and Rock (I). ASTM International.

## 3. Soils and Other Materials

- Cohesive soils
- Distilled water
- Microcrystalline wax (50 g)

## 4. Apparatus

- Balance of 0.01 g readability
- No. 40 sieve
- Drying oven maintaining a constant temperature of 110±5°C
- Thermometer of 0.1°C readability
- Shrinkage dish
- Porcelain dish of sufficient size to allow the soil pat to submerge
- Wax warmer
- Moisture content container with lid (2)
- Sewing thread – fine thread to hold the specimen to dip into the wax
- Spatula, straightedge, petroleum base lubricant

## 5. Preparation of Test Specimen

- Obtain 100 g of air-dry soil passing the No. 40 (425 μm) sieve.
- Mix the soil with distilled water using a spatula or straightedge. The amount of water added should produce water content above the liquid limit, based on visual inspection. The soil should flow sufficiently to expel air bubbles when using a gentle tapping action. It is desirable to use the minimum possible water to prevent cracks when drying.

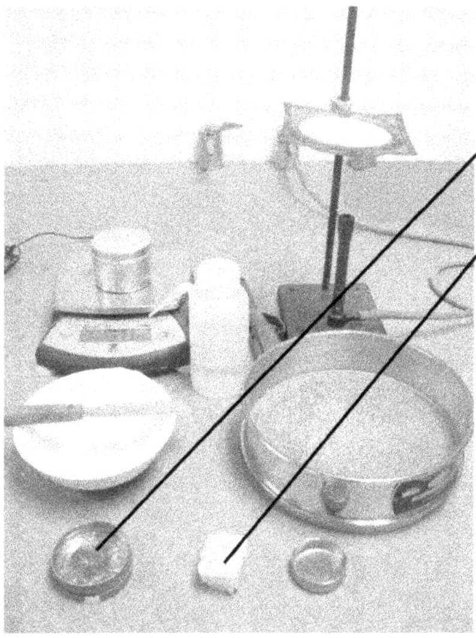


Figure 7-2. Apparatus and tools for shrinkage limit measurement.

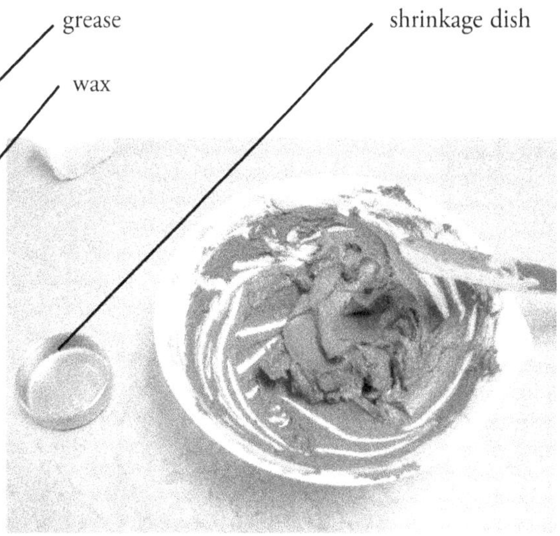


Figure 7-3. Soil preparation.

## 6. Test Procedure

(1) Measure and record the volume of the shrinkage dish, $V$.

(2) Lightly grease the inside of the shrinkage dish and determine the mass of the greased dish, $M_{dish}$.

(3) Fill the dish with soil:
- Place, in the center of the dish, an amount of wetted soil equal to approximately 1/3 of the volume of the dish. Gently tap the dish on a firm surface to cause the soil to flow to the edges until the dish is thoroughly compacted and all included air has been brought to the surface.
- Repeat the above step until the dish is completely filled and excess soil stands out above its edge.
- Strike off the excess soil using a straightedge and wipe off all soil adhering to the outside of the dish.

(4) Determine the mass of the shrinkage dish with soil, $M_{dish+wet\ soil}$.

(5) Allow the soil pat to dry in air until the color of the pat turns from dark to light. Then oven dry the pat at 110±5°C overnight. If the soil pat is cracked or has broken into pieces, clean the dish and return to Step (2) using soil with a lower water content.

(6) Determine the mass of the shrinkage dish with dry soil, $M_{dish+dry\ soil}$.

(7) Determine the volume of the dry soil pat using the wax method:
- Carefully remove the dry soil pat from the dish.
- Securely tie the sewing thread around the soil pat.
- Melt wax in a porcelain dish and immerse the dry soil pat in the melted wax to completely coat the pat. Do not allow air bubbles to develop in the wax coating. If air bubbles are present, use a sharp object to cut out the bubbles; then refill the holes with wax.
- Take out the soil pat from the melted wax and allow the wax coating to cool.
- Determine the mass of the dry soil pat with wax, $M_{dry\ soil+wax}$.
- Using the thread, suspend the soil pat from a balance and dip the coated soil pat into water. Measure the mass of the wax-coated dry soil in water, $M_{dry\ soil+wax\ (in\ water)}$.
- Measure the temperature of the water.

Figure 7-4. Soil pat before oven drying.

Figure 7-5. Soil pat after oven drying.

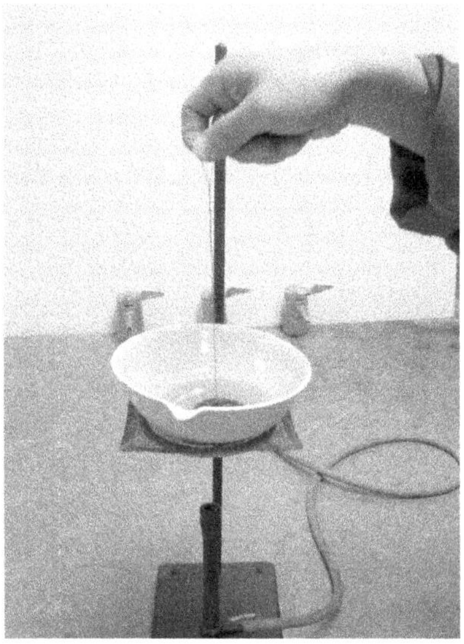

Figure 7-6. Coating the soil pat with wax.

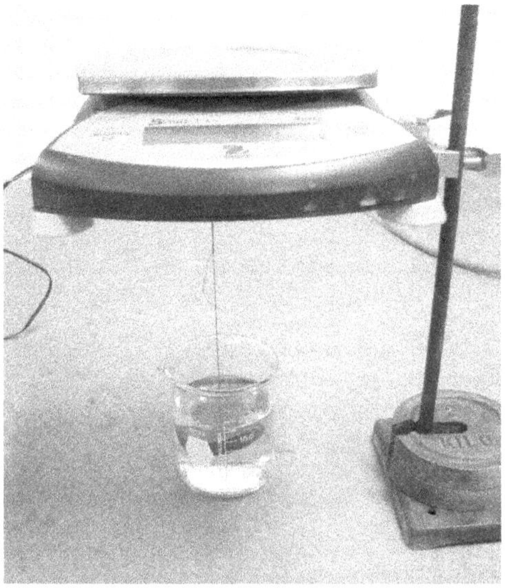

Figure 7-7. Measuring the soil pat mass in water.

# 7. Data Collection and Calculation

(a) Density of wax, $\rho_{wax}$ =

(b) Volume of the shrinkage dish, (= volume of the wet soil pat), $V$ =

(c) Mass of the shrinkage dish with grease, $M_{dish}$ =

(d) Mass of the shrinkage dish with wet soil, $M_{dish+wet\ soil}$ =

(e) Mass of the shrinkage dish with soil after oven drying, $M_{dish+dry\ soil}$ =

(f) Mass of the dry soil pat with wax (measured in air), $M_{dry\ soil+wax}$ =

(g) Mass of the dry soil pat with wax (measured in water), $M_{dry\ soil+wax\ (in\ water)}$ =

(h) Temperature of the water, T =

Density of the water at this temperature, $\rho_w$ =

Table 7-1. Water Density at Different Temperatures

| Temperature (°C) | Density (g/cm$^3$) | Temperature (°C) | Density (g/cm$^3$) |
|---|---|---|---|
| 15 | 0.999099 | 23 | 0.997538 |
| 16 | 0.998943 | 24 | 0.997296 |
| 17 | 0.998774 | 25 | 0.997044 |
| 18 | 0.998595 | 26 | 0.996783 |
| 19 | 0.998405 | 27 | 0.996512 |
| 20 | 0.998203 | 28 | 0.996232 |
| 21 | 0.997992 | 29 | 0.995944 |
| 22 | 0.997770 | 30 | 0.995646 |

## 8. Results and Analyses

(1) Calculate the mass of the dry soil pat:

$$M_{dry\_soil} = M_{dish+dry\_soil} - M_{dish}$$

(2) Calculate the initial water content:

$$w = \frac{M_{dish+wet\_soil} - M_{dish+dry\_soil}}{M_{dry\_soil}} \times 100\%$$

(3) Calculate the volume of the wax-coated dry soil pat, which is equal to the volume of water displaced by the soil pat. The mass of water displaced by the soil pat is equal to the mass of the soil pat in water.

$$V_{wax+dry\_soil} = V_{water\_displaced} = \frac{M_{dry\_soil+wax(in\_water)}}{\rho_w}$$

where:

$\rho_w$ = density of the water at the test temperature.

(4) Calculate the volume of wax coating:

$$V_{wax} = \frac{M_{wax}}{\rho_{wax}} = \frac{M_{wax+dry\_soil} - M_{dry\_soil}}{\rho_{wax}}$$

(5) Calculate the volume of the dry soil pat:

$$V_{dry\_soil} = V_{wax+dry\_soil} - V_{wax}$$

(6) Calculate the shrinkage limit (SL), which is the initial water content minus the water content loss due to the soil volume shrinkage as a result of oven-drying:

$$SL = w_{ini} - \frac{(V - V_{dry\_soil}) \cdot \rho_w}{M_{dry\_soil}} \times 100\%$$

## SAMPLE DATA COLLECTION, RESULTS, AND ANALYSES

(a) Density of wax, $\rho_{wax}$ = 0.90g/cm$^3$

(b) Volume of the shrinkage dish, (= volume of the wet soil pat), $V$ = 16.50 cm$^3$

(c) Mass of the shrinkage dish with grease, $M_{dish}$ = 24.21 g

(d) Mass of the shrinkage dish with wet soil, $M_{dish+wet\ soil}$ = 54.22 g

(e) Mass of the shrinkage dish with soil after oven drying, $M_{dish+dry\ soil}$ = 46.23 g

(f) Mass of the dry soil pat with wax (measured in air), $M_{dry\ soil+wax}$ = 22.71g

(g) Mass of the dry soil pat with wax (measured in water), $M_{dry\ soil+wax\ (in\ water)}$ = 10.12 g

(h) Temperature of the water, T = 21°C
Density of the water at this temperature, $\rho_w$ = 0.997992 g/cm$^3$

(1) Calculate the mass of the dry soil pat:

$$M_{dry\_soil} = M_{dish+dry\_soil} - M_{dish} = 46.23 - 24.21 = 22.02g$$

(2) Calculate the initial water content:

$$w_{ini} = \frac{M_{dish+wet\_soil} - M_{dish+dry\_soil}}{M_{dry\_soil}} \times 100\%$$

$$= \frac{54.22 - 46.23}{22.02} \times 100\% = 36.3\%$$

(3) Calculate the volume of the wax-coated dry soil pat, which is equal to the volume of water displaced by the soil pat. The mass of water displaced by the soil pat is equal to the mass of the soil pat in water.

$$V_{wax+dry\_soil} = V_{water\_displaced} = \frac{M_{dry\_soil+wax(in\_water)}}{\rho_w} = \frac{10.12}{0.997992} = 10.14cm^3$$

where:

$\rho_w$ = density of the water at the test temperature.

(4) Calculate the volume of wax coating:

$$V_{wax} = \frac{M_{wax}}{\rho_{wax}} = \frac{M_{dry\_soil+wax} - M_{dry\_soil}}{\rho_{wax}} = \frac{22.71 - 22.02}{0.90} = 0.77cm^3$$

(5) Calculate the volume of the dry soil pat:

$$V_{dry\_soil} = V_{wax+dry\_soil} - V_{wax} = 10.14 - 0.77 = 9.37cm^3$$

(6) Calculate the shrinkage limit (SL), which is the initial water content minus the water content loss due to the soil volume shrinkage as a result of oven-drying:

$$SL = w_{ini} - \frac{(V - V_{dry\_soil}) \cdot \rho_w}{M_{dry\_soil}} \times 100\%$$

$$= 36.3\% - \frac{(16.50 - 9.37) \times 0.997992}{22.02} \times 100\%$$

$$= 4.3\%$$

**The shrinkage limit is 4.3%.**

Name ______________________________

# Test 7: Shrinkage Limit of Cohesive Soils

Question 1: What is shrinkage limit?

Question 2: What are the particle sizes of a soil for shrinkage limit (SL) measurement?

Question 3: What are the two ASTM methods used for SL measurement?

# TEST 8

# Constant Head Permeability Test for Granular Soils

## 1. Introduction

The objective of this test is to determine the coefficient of permeability (hydraulic conductivity) of granular soils. The permeability, $k$, is calculated using Darcy's law:

$$v = k \cdot i$$

where:

$v$ = flow velocity in soil,

$$v = \frac{q}{A}$$

where:

$q$ = flow rate (volume of water passing a unit cross-sectional area per unit time), and
$A$ = cross-sectional area of the soil specimen,

$$q = \frac{Q}{t}$$

where:

$Q$ = the volume of water passing through the soil specimen during the time interval, $t$; and
$i$ = hydraulic gradient,

$$i = \frac{h}{L}$$

where:

$h$ = the total hydraulic head difference across the soil specimen of length, $L$.

Therefore, the coefficient of permeability can be calculated as follows:

$$k = \frac{QL}{Aht}$$

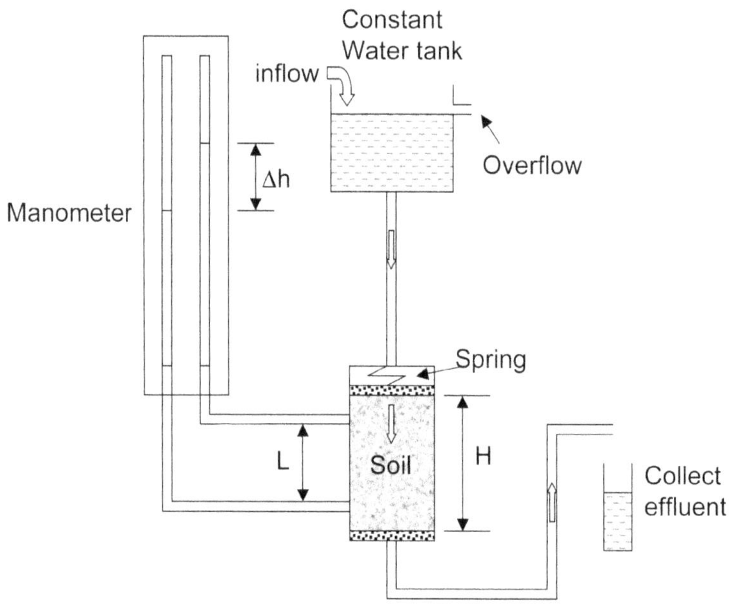


Figure 8-1. Constant head permeability test setup.

The test conditions that make Darcy's law valid are as follows:

(a) Soil volume does not change during the test.
(b) The soil specimen is fully saturated.
(c) The flow is steady state –there is no change in hydraulic gradient.
(d) The flow is laminar – the hydraulic gradient is small enough that the flow is not turbulent.

## 2. References

- ASTM D 2434-68 (Reapproved 2000) "Standard Test Methods for Permeability of Granular Soils," *2006 Annual Book of ASTM Standards*, Volume 04.08, Soil and Rock (I). ASTM International.

## 3. Soils and Other Materials

- Air-dried granular soils containing not more than 10% soil passing the No. 200 sieve
- De-aired water (optional)

## 4. Apparatus

- Permeameter, which includes (1) a clear specimen cylinder; (2) two porous disks or screens; (3) two manometer outlets for measuring the water head loss over a length, *L*, equivalent to at least the diameter of the cylinder; and (4) a spring attached to the top plate for applying a light spring pressure of 22 to 45 N (5 to 10 lbf) total load. The spring holds the specimen from changing in volume during the test.
- Constant-head tank
- Manometer tubes for measuring water head
- Vacuum pump capable of providing 50 cm (20 inches) Hg vacuum
- 250 mL graduated cylinder

- Sieve analysis equipment
- Specimen compaction equipment (such as vibrating tamper, sliding weight tamper, or other type of tamper)
- Large funnel used to pour soil into permeameter for compaction
- Water content measurement: drying oven, water content container
- Balance readable to 0.1 g
- Timer, caliper, thermometer, tubing, scoop, spray bottle

## 5. Preparation of Test Specimen

- Sieve analysis shall be made on a representative air-dried sample prior to the permeability test. Particles larger than 19 mm (3/4 inch) shall be separated out by sieving. The soil should contain no more than 10% of fines passing the No. 200 sieve.

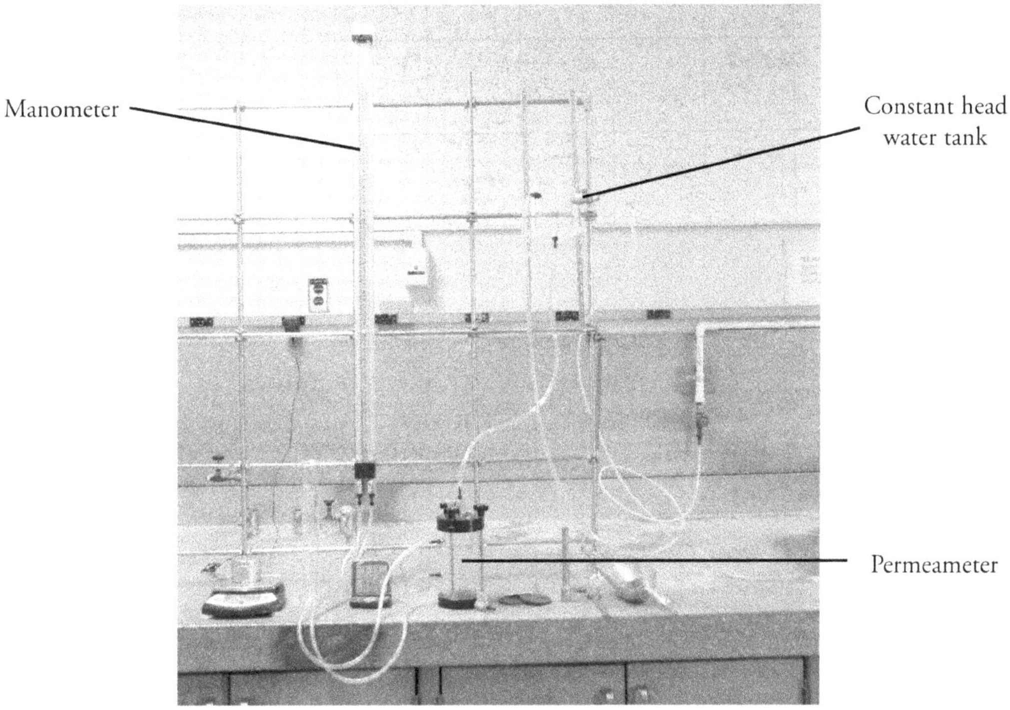


Figure 8-2. Constant head permeability test apparatus.

## 6. Test Procedure

(1) Measure the water content of the soil sample.
(2) Make the initial measurement of the permeameter cylinder:
  - The inside diameter, $D$
  - The length, $L$, between the two manometer outlets
(3) Prepare the soil specimen in the cylinder:
  - Specify the target density and the height of the soil specimen in the cylinder. The specimen should be above the upper manometer outlet by about 2.0 cm. Then the total mass of soil can be determined. Decide how many layers should be used in compaction and the target thickness and mass of each layer.
  - Place a porous disk or screen at the bottom of the cylinder.
  - Place one layer of soil into the cylinder using a scoop and a large funnel (Figure 8-3).

- Compaction: If minimum density is required, no compaction is needed; otherwise, compact each layer using a vibrating tamper, a sliding weight tamper, or other type of tamper.
- Repeat the compaction of subsequent layers.

Figure 8-3. Pouring sand into the permeameter.

Figure 8-4. Spring to hold the specimen.

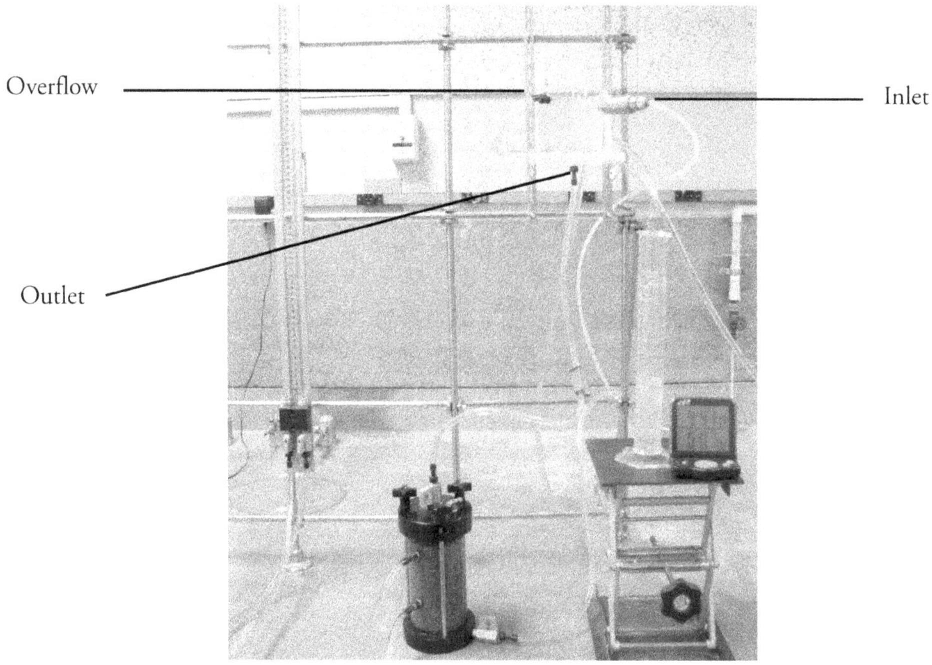


Figure 8-5. Permeability test in progress.

(4) Assemble the permeameter:
- After compaction, level the upper surface of the soil by placing the upper porous plate or screen in position and rotating it gently back and forth.
- Cap the cylinder with the top plate; the spring should be compressed to hold the specimen in place (Figure 8-4). Make sure the cap is sealed.
- Measure the height of the specimen, $H$, using a caliper at four symmetrically spaced points along the specimen. Calculate the average.
- Connect the water manometer tubes to the manometer outlets on the cylinder and clamp the tubing so that water does not flow into the manometer tubes during saturation.

(5) Saturating the specimen:
- Close the bottom specimen outlet and apply 50 cm (20 inches) Hg vacuum for 15 minutes through the top specimen outlet to remove some of the air from the specimen.
- Connect the bottom specimen outlet to the water tank with tubing to allow water to flow from the bottom upward. The tubing should be free of air before connecting to the outlet. The hydraulic gradient should be low enough to allow the water level to rise slowly in the specimen, taking no less than 10 minutes.
- When water emerges from the top outlet, close both outlets. Connect the top outlet to the upstream water tank, making sure the tubing is free of air. The bottom outlet can be open to the atmosphere or can be connected to tubing that is raised to a certain level to have tail water head.

(6) Conduct the permeability measurement:
- Fill the manometer tubes with water, making sure the manometer and the tubing are free of air. Open both outlet valves and adjust the input and output in the upstream water tank to achieve constant head.
- Suggested hydraulic gradient: on a loose-compaction specimen, 0.2 to 0.3; on a dense-compaction specimen, 0.3 to 0.5.
- Wait until the water levels in the manometer tubes are stable.
- Start the timer, and at the same time, start to collect effluent in a graduated cylinder. Measure and record the water levels in the two manometer tubes, the quantity of effluent, and the time.
- Measure the temperature of the effluent.

(7) Lower the downstream water level by 1 cm, and repeat Step (6).

(8) At the end of the test, drain the specimen and visually inspect whether the soil is homogeneous.

## 7. Data Collection and Calculation

Table 8-1. Water Content Measurement

| | |
|---|---|
| Mass of container, $M_1$ (g) | |
| Mass of container with moist soil specimen, $M_2$ (g) | |
| Mass of container with dry soil specimen, $M_3$ (g) | |
| Mass of water, $M_w = M_2 - M_3$ (g) | |
| Mass of dry soil, $M_s = M_3 - M_1$ (g) | |
| Moisture content, $w = \frac{M_w}{M_s} \times 100\%$ (%) | |

Table 8-2. Specimen Preparation and Compaction

| | |
|---|---|
| Specific gravity, $G_s$ | |
| Inside diameter of the cylinder, $D$ (cm) | |
| Specimen cross-sectional area, $A$ (cm$^2$) | |
| Target density of the specimen, $\rho_{target}$ (g/cm$^3$) | |
| Target specimen length, $L_{target}$ (cm) | |
| Mass of air-dried specimen needed, $M = \rho_{target} \cdot A \cdot L_{target}$ (g) | |
| Number of layers in compaction, $N$ | |
| Mass of soil per layer = $\frac{M}{N}$ (g) | |
| Thickness of each layer after compaction = $\frac{L_{target}}{N}$ (cm) | |
| Actual specimen length after compaction, $L_{actual}$ (cm) | |
| Actual specimen density after compaction, $\rho = \frac{M}{L_{actual} \cdot A}$ (g/cm$^3$) | |
| Dry density, $\rho_d = \frac{\rho}{1 + \frac{w(\%)}{100\%}}$ (g/cm$^3$) | |
| Void ratio, $e = \frac{G_s \rho_w}{\rho_d} - 1$ | |

Table 8-3. Permeability Measurements

Cross-sectional area of specimen, $A$ =

Sample length between the two manometer outlets, $L$ =

| | **Test 1** | **Test 2** |
|---|---|---|
| Time, $t$ (sec) | | |
| Volume of effluent collected, $Q$ (ml) | | |
| Water head in the upstream manometer tube, $h_1$ (cm) | | |
| Water head in the downstream manometer tube, $h_2$ (cm) | | |
| Water head difference, $h = h_1 - h_2$ (cm) | | |
| Temperature of effluent, $T$ (°C) | | |
| Hydraulic gradient, $h/L$ | | |
| Coefficient of permeability, $k = \dfrac{QL}{Aht}$ (cm/s) | | |

## 8. Results and Analyses

(1) Complete the calculations in Tables 8-1, 8-2, and 8-3.

(2) Calculate the average of the coefficient of permeability at the test temperature, $k_{T^0C}$ .

(3) Correct the permeability to that at 20°C (68°F) by multiplying $k_{T^0C}$ by the ratio of the viscosity of water at the test temperature, $\eta_{T^oC}$, to the viscosity of water at 20°C (68°F), $\eta_{20^oC}$.

$$k_{20^0C} = \frac{\eta_{T^0C}}{\eta_{20^0C}} k_{T^0C}$$

Table 8-4. Variation of $\dfrac{\eta_{T^0C}}{\eta_{20^0C}}$ with Temperature ($\eta_{20^oC}$ =1.0020 mPa·s)

| **Temperature, T (°C)** | $\dfrac{\eta_{T^0C}}{\eta_{20^0C}}$ | **Temperature, T (°C)** | $\dfrac{\eta_{T^0C}}{\eta_{20^0C}}$ |
|---|---|---|---|
| 15 | 1.135 | 23 | 0.931 |
| 16 | 1.106 | 24 | 0.910 |
| 17 | 1.077 | 25 | 0.889 |
| 18 | 1.051 | 26 | 0.869 |
| 19 | 1.025 | 27 | 0.850 |
| 20 | 1.000 | 28 | 0.832 |
| 21 | 0.976 | 29 | 0.814 |
| 22 | 0.953 | 30 | 0.797 |

## SAMPLE DATA COLLECTION, RESULTS, AND ANALYSES

Table 8-1. Water Content Measurement

| | |
|---|---|
| Mass of container, $M_1$ (g) | 27.3 |
| Mass of container with moist soil specimen, $M_2$ (g) | 129.4 |
| Mass of container with dry soil specimen, $M_3$ (g) | 128.9 |
| Mass of water, $M_w = M_2 - M_3$ (g) | 0.5 |
| Mass of dry soil, $M_s = M_3 - M_1$ (g) | 101.6 |
| Moisture content, $w = \frac{M_w}{M_s} \times 100\%$ (%) | 0.5 |

Table 8-2. Specimen Preparation and Compaction

| | |
|---|---|
| Specific gravity, $G_s$ | 2.65 |
| Inside diameter of the cylinder, $D$ (cm) | 7.626 |
| Specimen cross-sectional area, $A$ (cm$^2$) | 45.68 |
| Target density of the specimen, $\rho_{target}$ (g/cm$^3$) | 1.50 |
| Target specimen length, $L_{target}$ (cm) | 14.00 |
| Mass of air-dried specimen needed, $M = \rho_{target} \cdot A \cdot L_{target}$ (g) | 959.3 |
| Number of layers in compaction, $N$ | 4 |
| Mass of soil per layer = $\frac{M}{N}$ (g) | 239.8 |
| Thickness of each layer after compaction = $\frac{L_{target}}{N}$ (cm) | 3.5 |
| Actual specimen length after compaction, $L_{actual}$ (cm) | 13.83 |
| Actual specimen density after compaction, $\rho = \frac{M}{L_{actual} \cdot A}$ (g/cm$^3$) | 1.52 |
| Dry density, $\rho_d = \frac{\rho}{1 + \frac{w(\%)}{100\%}}$ (g/cm$^3$) | 1.51 |
| Void ratio, $e = \frac{G_s \rho_w}{\rho_d} - 1$ | 0.75 |

Table 8-3. Permeability Measurements

Cross-sectional area of specimen, $A$=45.68 $cm^2$
Sample length between the two manometer outlets, $L$ = 7.62cm

| | Test 1 | Test 2 |
|---|---|---|
| Time, $t$ (sec) | 152 | 195 |
| Volume of effluent collected, $Q$ (ml) | 250 | 230 |
| Water head in the upstream manometer tube, $h_1$ (cm) | 46.0 | 46.4 |
| Water head in the downstream manometer tube, $h_2$ (cm) | 42.4 | 43.8 |
| Water head difference, $h = h_1 - h_2$ (cm) | 3.6 | 2.6 |
| Temperature of effluent, $T$ (°C) | 21.0 | 21.0 |
| Hydraulic gradient, $h/L$ | 0.47 | 0.34 |
| Coefficient of permeability, $k = \frac{QL}{Aht}$ (cm/s) | $7.60\times10^{-2}$ | $7.57\times10^{-2}$ |

The average of the coefficient of permeability: $k_{21^0C} = 7.58 \times 10^{-2} cm/\sec$

Correct the permeability to that at 20°C (68°F) by multiplying $k$ by the ratio of the viscosity of water at the test temperature, $\eta_{T^oC}$, to the viscosity of water at 20°C (68°F), $\eta_{20^oC}$.

From Table 8-4: at 21°C, $\frac{\eta_{21^0C}}{\eta_{20^0C}} = 0.976$

So: $k_{20^0C} = \frac{\eta_{T^0C}}{\eta_{20^0C}} k_{T^0C} = 0.976 \times (7.58 \times 10^{-2}) = 7.40 \times 10^{-2} cm/s$

Name ______________________________

# Test 8: Constant Head Permeability Test for Granular Soils

Question 1: List the Darcy's Law and define the terms in the equation.

Question 2: What is the range of the ASTM-suggested hydraulic gradient of a soil specimen in a constant head permeability test?

Question 3: A sandy soil is tested for permeability using the constant head method. The diameter of the soil column is 8.0 cm, the length of the soil column is 16.0 cm. The hydraulic head loss across the soil specimen is 6.4 cm. Over a period of 3 min, 350 ml of water is collected at the downstream side of the soil specimen. What is the permeability of the soil specimen (cm/sec)?

TEST 9

# Falling Head Permeability Test for Fine-Grained Soils

## 1. Introduction

The objective of this test is to determine the coefficient of permeability (hydraulic conductivity) of fine-grained soils, although the method also applies to granular soils. The falling head method is used primarily for economy, since a test to determine permeability for fine-grained soil on the order of $10^{-4}$ to $10^{-9}$ cm/sec may take many hours or days. To maintain a constant head for fine-grained soil over long duration, it may waste a large quantity of water as the result of overflow.

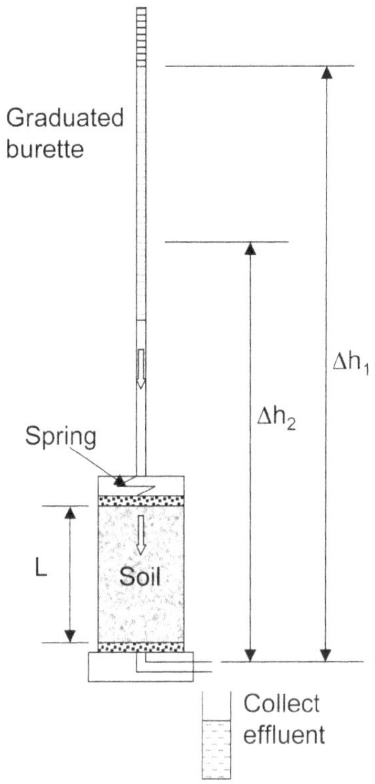


Figure 9-1. Falling head permeability test setup.

The permeability, $k$, is calculated based on Darcy's law:

$$v = k \cdot i$$

where:
$v$ = flow velocity in soil, and
$i$ = hydraulic gradient.

Since the hydraulic head difference across a soil specimen decreases with time, the permeability can be derived by integrating a differential equation in terms of time and hydraulic head difference and is expressed as follows:

$$k = \frac{aL}{At}\ln\frac{\Delta h_1}{\Delta h_2} = 2.303\frac{aL}{At}\log_{10}\frac{\Delta h_1}{\Delta h_2}$$

where:

$a$ = cross-sectional area of burette or other standpipe connected to the upstream of the soil specimen,
$A$ = cross-sectional area of the soil specimen,
$L$ = length of the soil specimen,
$\Delta h_1$ = hydraulic head difference across the specimen at the beginning of the test ($t$ = 0),
$\Delta h_2$ = hydraulic head difference across the specimen at the end of the test ($t$>0), and
$t$ = duration of the test.

The test conditions that make Darcy's law valid are as follows:
(a) Soil volume does not change during the test.
(b) The soil specimen is fully saturated.
(c) The flow is steady state – there is no change in hydraulic gradient.
(d) The flow is laminar – the hydraulic gradient is small enough that the flow is not turbulent.

One disadvantage of using the falling head method for fine-grained soils is the possible sidewall leakage – water flows between the soil specimen and the smooth rigid wall. For accurate measurement, the flexible wall method using triaxial apparatus (ASTM D5084) is recommended.

## 2. References

This test has not been standardized by ASTM as of 2008.

## 3. Soils and Other Materials

- Fine-grained soils
- De-aired water (optional)

## 4. Apparatus

- Permeameter, which includes (1) a clear specimen cylinder; (2) two porous disks or screens; and (3) a spring attached to the top plate for applying a light spring pressure of 22 to 45 N (5 to 10 lbf) total load. The spring holds the specimen from changing in volume during the test.

- Graduated burette of standpipe
- Vacuum pump capable of providing 50 cm (20 inches) Hg vacuum
- Specimen compaction equipment (such as vibrating tamper, sliding weight tamper, or other type of tamper)
- Large funnel used to pour soil into the permeameter for compaction
- Water content measurement: drying oven, water content container
- Balance readable to 0.1 g
- Timer, caliper, thermometer, tubing, scoop, spray bottle

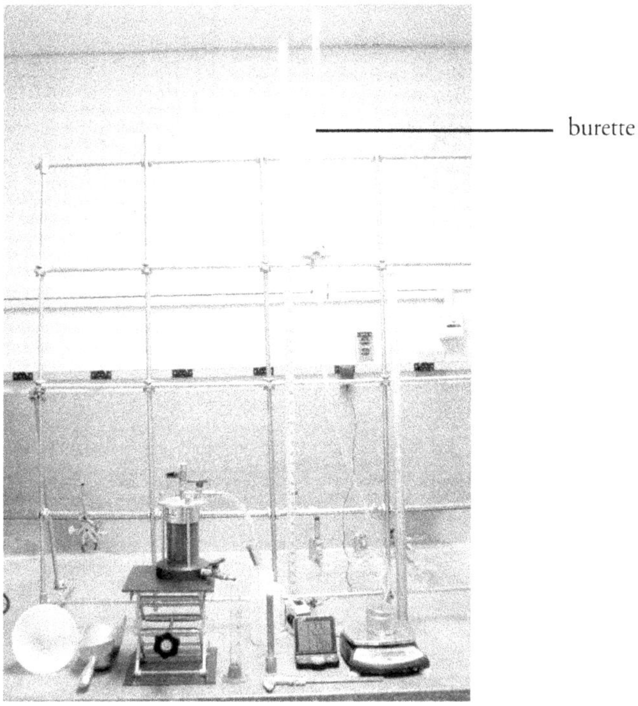


Figure 9-2. Falling head permeability test apparatus and setup.

## 5. Preparation of Test Specimen

- The soil is compacted in the permeameter as described in the Test Procedure.

## 6. Test Procedure

(1) Measure the water content of the soil sample.
(2) Measure the inside diameter, $D$, of the permeameter cylinder.
(3) Prepare the soil specimen in the cylinder:
- Specify the target density and the target length of the soil specimen in the cylinder. Then the total mass of soil can be determined. Decide how many layers should be used in compaction and the target thickness and mass of each layer.
- Place a porous disk or screen at the bottom of the cylinder.
- Place one layer of soil into the cylinder using a scoop and a large funnel.
- Compaction: If minimum density is required, no compaction is needed; otherwise, compact each layer using a vibrating tamper, or a sliding weight tamper, or another type of tamper.
- Repeat the compaction of subsequent layers.

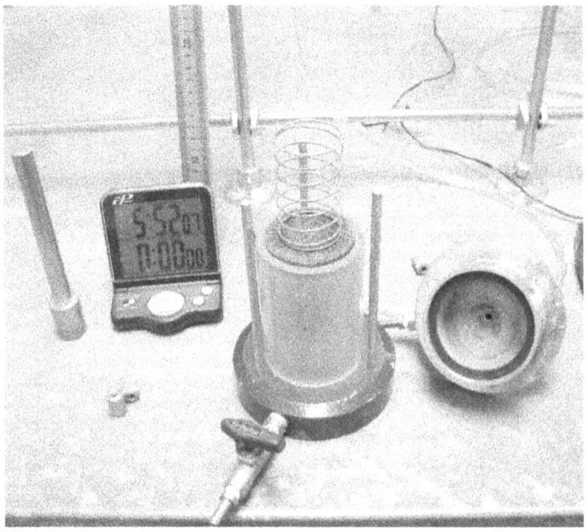

Figure 9-3. Spring to hold specimen.

Figure 9-4. Saturation process.

(4) Assemble the permeameter:
- After compaction, level the upper surface of the soil by placing the upper porous plate or screen in position and rotating it gently back and forth.
- Cap the cylinder with the top plate; the spring should be compressed to hold the specimen in place. Make sure the cap is sealed.
- Measure the height of the specimen, $L$, using a caliper at four symmetri-cally spaced points along the specimen. Calculate the average.

(5) Saturating the specimen:
- Close the bottom outlet, and apply 50 cm (20 inches) Hg vacuum for 15 minutes through the top outlet to remove some of the air from the specimen.
- Connect the bottom outlet to the burette or standpipe using tubing to allow water to flow from the bottom upward. The tubing should be free of air before connecting to the outlet. The hydraulic gradient should be low enough to allow the water level to rise slowly in the specimen, taking no less than 10 minutes.
- When water emerges from the top outlet, close both outlets. Disconnect the burette from the bottom outlet and connect the burette to the top outlet, making sure the tubing is free of air. The bottom outlet can be open to the atmosphere. (This is step is not needed if the flow is upward during the permeability measurement.)

(6) Conduct the permeability measurement:
- Open both outlet valves of the permeameter, and allow water to run through the specimen for a few minutes to reach a stable flow.
- Close both outlet valves. Add water to the burette to a level near the top of the burette. Record the initial reading of the burette, $V_1$. Measure the height from the effluent outlet to the bottom of the meniscus in the burette; this is the initial water head difference, $\Delta h_1$.
- Open both outlet valves and start the timer. Collect effluent in a graduated cylinder. After the water level drops significantly, close both valves and stop the timer at the same time. Measure the temperature of the effluent. Record the final reading of the burette, $V_2$. Measure the height from the effluent outlet to the bottom of the menis-

cus in the burette; this is the final water head difference, $\Delta h_2$. Record the time. Repeat this step to obtain a second measurement.

(7) At the end of the test, drain the specimen and visually inspect whether the soil is homogeneous.

## 7. Data Collection and Calculation

Table 9-1. Water Content Measurement

| | |
|---|---|
| Mass of container, $M_1$ (g) | |
| Mass of container with moist soil specimen, $M_2$ (g) | |
| Mass of container with dry soil specimen, $M_3$ (g) | |
| Mass of water, $M_w = M_2 - M_3$ (g) | |
| Mass of dry soil, $M_s = M_3 - M_1$ (g) | |
| Moisture content, $w = \frac{M_w}{M_s} \times 100\%$ (%) | |

Table 9-2. Specimen Preparation and Compaction

| | |
|---|---|
| Specific gravity, $G_s$ | |
| Inside diameter of the cylinder, $D$ (cm) | |
| Specimen cross-sectional area, $A$ (cm$^2$) | |
| Target density of specimen, $\rho_{target}$ (g/cm$^3$) | |
| Target specimen length, $L_{target}$ (cm) | |
| Mass of air-dried specimen needed, $M = \rho_{target} \cdot A \cdot L_{target}$ (g) | |
| Number of layers in compaction, $N$ | |
| Mass of soil per layer = $\frac{M}{N}$ (g) | |
| Thickness of each layer after compaction = $\frac{L_{target}}{N}$ (cm) | |
| Actual specimen length after compaction, $L_{actual}$ (cm) | |
| Actual specimen density after compaction, $\rho = \frac{M}{L_{actual} \cdot A}$ (g/cm$^3$) | |
| Dry density, $\rho_d = \frac{\rho}{1 + \frac{w(\%)}{100\%}}$ (g/cm$^3$) | |
| Void ratio, $e = \frac{G_s \rho_w}{\rho_d} - 1$ | |

Table 9-3. Permeability Measurements

Cross-sectional area of the specimen (from Table 9-2), $A$ =
Length of specimen (from Table 9-2), $L_{actual}$ =

| | **Test 1** | **Test 2** |
|---|---|---|
| Initial burette reading, $V_1$ (ml) | | |
| Initial water head difference, $\Delta h_1$ (cm) | | |
| Final burette reading, $V_2$ (ml) | | |
| Final water head difference, $\Delta h_2$ (cm) | | |
| Time, $t$ (sec) | | |
| Temperature of the effluent, $T$ (°C) | | |
| Cross-sectional area of the burette, $a = \dfrac{\lvert V_1 - V_2 \rvert}{h_1 - h_2}$ (cm$^2$) | | |
| Average cross-sectional area of the burette (cm$^2$) | | |
| Coefficient of permeability, $k = \dfrac{aL}{At}\ln\dfrac{\Delta h_1}{\Delta h_2} = 2.303\dfrac{aL}{At}\log_{10}\dfrac{\Delta h_1}{\Delta h_2}$ (cm/s) | | |

## 8. Results and Analyses

(1) Complete the calculations in Tables 9-1, 9-2, and 9-3.

(2) Calculate the average of the coefficient of permeability at the test temperature, $k_{T^0C}$.

(3) Correct the permeability to that at 20°C (68°F) by multiplying $k_{T^0C}$ by the ratio of the viscosity of water at the test temperature, $\eta_{T^oC}$, to the viscosity of water at 20°C (68°F), $\eta_{20^oC}$.

$$k_{20^0C} = \frac{\eta_{T^0C}}{\eta_{20^0C}} k_{T^0C}$$

Variation of $\dfrac{\eta_{T^0C}}{\eta_{20^0C}}$ with temperature can be found in Table 8-4 of Test 8.

# SAMPLE DATA COLLECTION, RESULTS, AND ANALYSES

Table 9-1. Water Content Measurement

| | |
|---|---|
| Mass of container, $M_1$ (g) | 27.0 |
| Mass of container with moist soil specimen, $M_2$ (g) | 124.3 |
| Mass of container with dry soil specimen, $M_3$ (g) | 123.1 |
| Mass of water, $M_w = M_2 - M_3$ (g) | 1.2 |
| Mass of dry soil, $M_s = M_3 - M_1$ (g) | 96.1 |
| Moisture content, $w = \dfrac{M_w}{M_s} \times 100\%$ (%) | 1.2 |

Table 9-2. Specimen Preparation and Compaction

| | |
|---|---|
| Specific gravity, $G_s$ | 2.65 |
| Inside diameter of the cylinder, $D$ (cm) | 6.267 |
| Specimen cross-sectional area, $A$ (cm$^2$) | 30.85 |
| Target density of specimen, $\rho_{target}$ (g/cm$^3$) | 1.6 |
| Target specimen length, $L_{target}$ (cm) | 8.0 |
| Mass of air-dried specimen needed, $M = \rho_{target} \cdot A \cdot L_{target}$ (g) | 394.8 |
| Number of layers in compaction, $N$ | 4 |
| Mass of soil per layer = $\dfrac{M}{N}$ (g) | 98.7 |
| Thickness of each layer after compaction = $\dfrac{L_{target}}{N}$ (cm) | 2.0 |
| Actual specimen length after compaction, $L_{actual}$ (cm) | 7.60 |
| Actual specimen density after compaction, $\rho = \dfrac{M}{L_{actual} \cdot A}$ (g/cm$^3$) | 1.68 |
| Dry density, $\rho_d = \dfrac{\rho}{1 + \dfrac{w(\%)}{100\%}}$ (g/cm$^3$) | 1.66 |
| Void ratio, $e = \dfrac{G_s \rho_w}{\rho_d} - 1$ | 0.59 |

Table 9-3. Permeability Measurements

Cross-sectional area of the specimen (from Table 9-2), $A$ = 30.85 cm$^2$
Length of specimen (from Table 9-2), $L = L_{actual}$ = 7.60 cm

| | Test 1 | Test 2 |
|---|---|---|
| Initial burette reading, $V_1$ (ml) | 3.8 | 34.8 |
| Initial water head difference, $\Delta h_1$ (cm) | 121.8 | 103.5 |
| Final burette reading, $V_2$ (ml) | 34.8 | 80.0 |
| Final water head difference, $\Delta h_2$ (cm) | 103.5 | 76.7 |
| Time, $t$ (sec) | 480 | 900 |
| Temperature of the effluent, $T$ (°C) | 21.0 | 21.0 |
| Cross-sectional area of the burette, $a = \frac{\lvert V_1 - V_2 \rvert}{h_1 - h_2}$ (cm$^2$) | 1.694 | 1.687 |
| Average cross-sectional area of the burette (cm$^2$) | 1.690 | |
| Coefficient of permeability, $k = \frac{aL}{At}\ln\frac{\Delta h_1}{\Delta h_2} = 2.303\frac{aL}{At}\log_{10}\frac{\Delta h_1}{\Delta h_2}$ (cm/s) | $1.41\times10^{-4}$ | $1.39\times10^{-4}$ |

(1) Complete the calculations in Tables 9-1, 9-2, and 9-3.

(2) Calculate the average of the coefficient of permeability at the test temperature, $k_{T^0C}$.

(3) Correct the permeability to that at 20°C (68°F) by multiplying $k_{T^0C}$ by the ratio of the viscosity of water at the test temperature, $\eta_{T^oC}$, to the viscosity of water at 20°C (68°F), $\eta_{20^oC}$.

$$k_{20^0C} = \frac{\eta_{T^0C}}{\eta_{20^0C}} k_{T^0C}$$

The average of the coefficient of permeability $k_{21^0C} = 1.40 \times 10^{-4} cm/\sec$:

Correct the permeability to that at 20°C (68°F) by multiplying $k_{21^0C}$ by the ratio of the viscosity of water at the test temperature, $\eta_{21^oC}$, to the viscosity of water at 20°C (68°F), $\eta_{20^oC}$.

From Table 8-4 (in Test 8): at 21°C, $\frac{\eta_{21^0C}}{\eta_{20^0C}} = 0.976$ ,

So: $k_{20^0C} = \frac{\eta_{T^0C}}{\eta_{20^0C}} k_{T^0C} = 0.976 \times (1.40 \times 10^{-4}) = 1.37 \times 10^{-4} cm/s$

Name ______________________________

# Test 9: Falling Head Permeability Test for Find-Grained Soils

Question 1: List the Darcy's Law and define the terms in the equation.

Question 2: A silty soil is tested for permeability using the falling head method. The diameter of the soil column is 8.0 cm, the length of the soil column is 16.0 cm. The hydraulic head loss across the soil specimen is 120.0 cm at the beginning of the test; at the end of 15 min, the hydraulic head loss across the soil specimen is 90.0 cm. The cross-sectional area of the burette that connects to the upstream of the soil specimen is 2.0 $cm^2$. What is the permeability of the soil specimen (cm/sec)?

# TEST 10

# Consolidation Test Using Incremental Loading

## 1. Introduction

The objective of this test is to determine the magnitude and the rate of consolidation of soil when it is restrained laterally and drained axially while subjected to incrementally applied controlled-stress loading. Each stress increment is maintained until excessive pore water pressure is completely dissipated. During the consolidation process, the changes in specimen height are measured and used to determine the relationship between the effective stress and the void ratio or axial strain and the rate at which consolidation can occur by evaluating the coefficient of consolidation.

## 2. References

- ASTM D 2435-04 "Standard Test Methods for One-Dimensional Consolidation Properties of Soils Using Incremental Loading," *2006 Annual Book of ASTM Standards*, Volume 04.08, Soil and Rock (I). ASTM International.
- ASTM D 4186-89 (Reapproved 1998) "Standard Test Methods for One-Dimensional Consolidation Properties of Soils Using Controlled-Strain Loading," *2006 Annual Book of ASTM Standards*, Volume 04.08, Soil and Rock (I). ASTM International.

## 3. Soils and Other Materials

- The test is most commonly performed on undisturbed samples of *fine-grained* soils
- Distilled or demineralized water

## 4. Apparatus

- Consolidometer and load device with weights
- Porous disks (2) and filter paper (2)
- Deformation indicator, such as vertical displacement dial gauge or linear variable differential transformer (LVDT) transducer with readout
- Trimming device such as cylindrical cutting ring (same inside diameter as the consolidometer ring) or trimming turntable with wire saw
- Balance of 0.01 g readability
- Drying oven maintaining a constant temperature of 110±5°C
- Water content containers (2)
- Clock, straightedge, screwdriver

## 5. Consolidometer Calibration

Due to the compressibility of porous disks, filter paper, and the consolidometer, the measured vertical deformation in the test must be corrected:

Figure 10-1. Consolidation test apparatus.

(1) Select a copper or hard-steel disk of the same height as the soil specimen and a slightly smaller diameter than the ring. Moisten the porous stone and the filter paper.
(2) Assemble the consolidometer with porous stones, filter paper, and disk.
(3) Load and unload the consolidometer as will be performed in the consolidation test. Wait for 2 minutes and measure the deformation at each load applied.
(4) Plot or tabulate the corrections at each load applied. The results can be incorporated in Table 10-5.

## 6. Preparation of Test Specimen

- Soil specimens are from undisturbed soils collected using sampling tubes. The samples obtained should be larger than or equal to the inside diameter of the consolidometer ring.

## 7. Test Procedure

(1) Measure the mass of the consolidometer ring, $M_r$.
(2) Trim the specimen and insert it into the consolidometer ring:

Figure 10-2. Specimen preparation.

Figure 10-3. Consolidation in progress.

- Trim the soil specimen using a cylindrical cutting ring or a trimming turntable with a wire saw to the same diameter as the consolidometer ring.
- Carefully insert the specimen into the consolidometer ring with a minimum of force, until the specimen protrudes from the ring.
- Trim the specimen flush with the plane ends of the ring using a wire saw and straightedge.

(3) Measure the mass of the consolidometer ring with the soil specimen, $M_{rs}$. Measure the diameter ($D$) and height ($H$) of the specimen.

(4) Measure the initial water content of the specimen using the remaining soil.

(5) Assemble the consolidometer on the loading device. In this test demonstration, a fixed-ring consolidometer is used. The consolidometer sits on a rigid base (Figure 10-2).
- Use porous stones at the bottom and the top of the specimen. Filter papers are preferred to prevent clogging of the porous stones.
  - If the soil is dry, dry filter papers and porous stones should be used.
  - If the soil is damp, damp filter papers and porous stones should be used.
  - If the soil is saturated, saturated filter papers and porous stones should be used.
- Ensure the ring is sealed with the base to prevent water leakage.

(6) After the consolidometer is placed on the loading device, apply a seating pressure of 5kPa (100 lbf/ft$^2$). If it is anticipated that a load of 5kPa will cause significant consolidation of the specimen, reduce the seating pressure to 2.5kPa or less. Tare the vertical displacement dial gauge.

(7) Inundate the specimen (add water to cover the specimen).

(8) Apply incremental loads to the specimen to obtain stresses of 12, 25, 50, 100, 200, etc., kPa. The standard load increment duration is 24 hours. Take the deformation reading at the end of each load, before doubling the load.

(9) For at least two load increments, including at least one load larger than the preconsolidation pressure, record the vertical displacement dial gauge readings at 0.1, 0.25, 0.5, 1, 2, 4, 8, 15, and 30 minutes and at 1, 2, 4, 8, and 24 hours. The data are used to obtain the coefficient of consolidation.

(10) If unload-reload is required, at least two decrements of pressure are needed. The duration of each unloading and reloading is 24 hours.

(11) At the end of the last loading, take the load off and disassemble the consolidometer. Measure the mass of the consolidometer ring and the specimen. Measure the water content of the final specimen.

## 8. Data Collection and Calculation

Table 10-1. Initial Moisture Content of the Specimen

| | |
|---|---|
| Mass of moisture can (g) | |
| Mass of moisture can and specimen (g) | |
| Mass of moisture can and specimen soil after oven-drying (g) | |
| Dry soil mass (g) | |
| Water in the soil (g) | |
| Water content, $w_{ini}$ (%) | |

Table 10-2. Specimen Dimensions

| | |
|---|---|
| Specimen diameter, $D$ (cm) | |
| Cross-sectional area of the specimen, $A$ (cm$^2$) | |
| Initial specimen height, $H_0$ (cm) | |
| Initial volume of specimen, $V_t$ (cm$^3$) | |

Table 10-3. Initial Mass and Density of the Specimen

| | |
|---|---|
| Mass of consolidometer ring, $M_r$ (g) | |
| Mass of consolidometer ring and specimen, $M_{rs}$ (g) | |
| Mass of specimen, $M_s = M_{rs} - M_r$ (g) | |
| Mass of dry specimen, $M_d = \dfrac{M_s}{1+\dfrac{w_{ini}(\%)}{100\%}}$ (g) | |
| Initial density of the specimen, $\rho_0 = \dfrac{M_s}{V_t}$ (g/cm3) | |
| Initial dry density of the specimen, $\rho_d = \dfrac{M_d}{V_t}$ (g/cm$^3$) | |

**Void Ratio Calculation:**

$$e = \frac{V_v}{V_s} = \frac{V_t - V_s}{V_s} = \frac{\frac{V_t}{A} - \frac{V_s}{A}}{\frac{V_s}{A}} = \frac{H_t - H_s}{H_s}$$

where:

$A$ = cross-sectional area of specimen,
$H_t$ = total height of the specimen (it changes with load increment),
$H_s = \frac{V_s}{A}$, equivalent height of soil solids; $H_s$ is assumed to remain constant,
$V_s$ = volume of soil solids,

$$V_s = \frac{M_d}{\rho_w G_s}$$

where:

$M_d$ = dry soil mass,
$G_s$ = specific gravity of soil solids.

Table 10-4. Equivalent Height of Soil Solids

| | |
|---|---|
| Specific gravity, $G_s$ | |
| Mass of dry specimen (From Table 10-3), $M_d$ (g) | |
| Volume of soil solids, $V_s$ (cm$^3$) | |
| Equivalent height of solids, $H_s$ (cm) | |

Table 10-5. Void Ratio and Normal Stress During Consolidation

| Mass of Total Loading Plates (kg) | Load on the Specimen (kN) | Normal Pressure (kN/m$^2$) | Dial Reading without Specimen* $\Delta H_C$ | Final Dial Reading** $R$ | Height Decrement $\Delta H = R - \Delta H_C$ | Final Specimen Height $H_t = H_0 - \Delta H$ | Void Ratio $e = \frac{H_t - H_s}{H_s}$ |
|---|---|---|---|---|---|---|---|
| | | | | | | | |
| | | | | | | | |
| | | | | | | | |
| | | | | | | | |
| | | | | | | | |
| | | | | | | | |
| | | | | | | | |

* Consolidometer calibration
** Height decrement of specimen at the **end** of each load

Table 10-6. Vertical Dial Reading vs. Time

Normal pressure =

Consolidometer correction for the normal pressure above, $\Delta H_C$ =

| **Time after Load Application, *t* (min)** | $\sqrt{t}$ **(min$^{0.5}$)** | **Vertical Dial Reading *R*** | **Vertical Dial Reading after Correction $R\text{–}\Delta H_C$** |
|---|---|---|---|
| 0.1 (6 sec) | | | |
| 0.25 (15 sec) | | | |
| 0.5 (30 sec) | | | |
| 1 | | | |
| 2 | | | |
| 4 | | | |
| 8 | | | |
| 15 | | | |
| 30 | | | |
| 60 (1 hr) | | | |
| 120 (2 hr) | | | |
| 240 (4hr) | | | |
| 480 (8hr) | | | |
| 1440 (24hr) | | | |

Table 10-7. Final Moisture Content of the Specimen

| | |
|---|---|
| Mass of moisture can (g) | |
| Mass of moisture can and specimen (g) | |
| Mass of moisture can and specimen soil after oven-drying (g) | |
| Dry soil mass (g) | |
| Water in the soil (g) | |
| Water content, *w* (%) | |

## 9. Results and Analyses

(1) Complete the calculations of the tables above.

(2) From Table 10-5, Draw an *e*-log*p* curve to evaluate the preconsolidation pressure, $\sigma'_c$, and the compression index, $c_c$. $\sigma'_c$ is graphically obtained using the Casagrande method; $c_c$ is the slope of the straight line portion of the *e*-log*p* curve.

(3) From Table 10-6, draw the time-deformation curve. It is used to predict the completion of the primary consolidation and the coefficient of secondary consolidation.

(4) From Table 10-6, draw the $\sqrt{time}$ -deformation curve using the square-root-of-time method. It is used to evaluate the coefficient of consolidation.
The graphs are illustrated in the sample section.

## SAMPLE DATA COLLECTION, CALCULATION, RESULTS, AND ANALYSES

Table 10-1. Initial Moisture Content of the Specimen

| | |
|---|---|
| Mass of moisture can (g) | 27.0 |
| Mass of moisture can and specimen (g) | 107.1 |
| Mass of moisture can and specimen soil after oven-drying (g) | 86.7 |
| Dry soil mass (g) | 59.7 |
| Water in the soil (g) | 20.4 |
| Water content, $w_{ini}$ (%) | 34.1 |

Table 10-2. Specimen Dimensions

| | |
|---|---|
| Specimen diameter, $D$ (cm) | 6.35 |
| Cross-sectional area of the specimen, $A$ (cm$^2$) | 31.67 |
| Initial specimen height, $H_0$ (cm) | 2.54cm = 1.0in |
| Initial volume of specimen, $V_t$ (cm$^3$) | 80.44 |

Table 10-3. Initial Mass and Density of the Specimen

| | |
|---|---|
| Mass of consolidometer ring, $M_r$ (g) | 537.1 |
| Mass of consolidometer ring and specimen, $M_{rs}$ (g) | 683.1 |
| Mass of specimen, $M_s = M_{rs} - M_r$ (g) | 146.0 |
| Mass of dry specimen, $M_d = \dfrac{M_s}{1 + \dfrac{w_{ini}(\%)}{100\%}}$ (g) | 108.9 |
| Initial density of the specimen, $\rho_0 = \dfrac{M_s}{V_t}$ (g/cm3) | 1.82 |
| Initial dry density of the specimen, $\rho_d = \dfrac{M_d}{V_t}$ (g/cm$^3$) | 1.35 |

Table 10-4. Equivalent Height of Solids

| | |
|---|---|
| Specific gravity, $G_s$ | 2.66 |
| Mass of dry specimen (From Table 10-3), $M_d$ (g) | 108.9 |
| Volume of soil solids, $V_s$ (cm$^3$) | 40.94 |
| Equivalent height of solids, $H_s$ (cm) | 1.29cm = 0.5079in |

Table 10-5. Void Ratio and Normal Stress During Consolidation
(The precision of the dial gauge is 0.0001 inch)

| **Mass of Total Loading Plates (kg)** | **Load on the Specimen (kN)** | **Normal Pressure (kN/m$^2$)** | **Dial Reading without Specimen* $\Delta H_C$ (in)** | **Final Dial Reading** $R$ (in)** | **Height Decrement $\Delta H = R - \Delta H_C$ (in)** | **Final Specimen Height (in) $H_t = H_0 - \Delta H$** | **Void Ratio $e = \frac{H_t - H_s}{H_s}$** |
|---|---|---|---|---|---|---|---|
| 1 | 0.0491 | 15.50 | 0.0001 | 0.0557 | 0.0556 | 0.9444 | 0.859 |
| 2 | 0.0981 | 30.97 | 0.0004 | 0.0698 | 0.0694 | 0.9306 | 0.832 |
| 4 | 0.1962 | 61.95 | 0.0011 | 0.0977 | 0.0957 | 0.9043 | 0.780 |
| 8 | 0.3924 | 123.90 | 0.0024 | 0.1320 | 0.1296 | 0.8704 | 0.714 |
| 16 | 0.7848 | 247.80 | 0.0043 | 0.1683 | 0.1640 | 0.8360 | 0.646 |

* Consolidometer calibration

** Height decrement of specimen at the end of each load

Table 10-6. Vertical Dial Reading vs. Time

Normal pressure = 61.95 kN/m²
Consolidometer correction, $\Delta H_C$ = 0.0011in

| **Time after Load Application, *t* (min)** | $\sqrt{t}$ **(min$^{0.5}$)** | **Vertical Dial Reading *R* (in)** | **Vertical Dial Reading after Correction $R\text{–}\Delta H_C$ (in)** |
|---|---|---|---|
| 0.1 (6 sec) | 0.32 | 0.0051 | 0.0040 |
| 0.25 (15 sec) | 0.50 | 0.0060 | 0.0049 |
| 0.5 (30 sec) | 0.71 | 0.0072 | 0.0061 |
| 1 | 1.00 | 0.0086 | 0.0075 |
| 2 | 1.41 | 0.0108 | 0.0097 |
| 4 | 2.00 | 0.0130 | 0.0119 |
| 8 | 2.83 | 0.0159 | 0.0148 |
| 15 | 3.87 | 0.0184 | 0.0173 |
| 30 | 5.48 | 0.0211 | 0.0200 |
| 60 (1 hr) | 7.75 | 0.0230 | 0.0219 |
| 120 (2 hr) | 10.95 | 0.0248 | 0.0237 |
| 240 (4 hr) | 15.49 | 0.0260 | 0.0249 |
| 480 (8 hr) | 21.91 | 0.0270 | 0.0259 |
| 1440 (24 hr) | 37.95 | 0.0279 | 0.0268 |

Table 10-7. Final Moisture Content of the Specimen

| | |
|---|---|
| Mass of moisture can (g) | 27.7 |
| Mass of moisture can and specimen (g) | 160.6 |
| Mass of moisture can and specimen soil after oven-drying (g) | 135.3 |
| Dry soil mass (g) | 107.6 |
| Water in the soil (g) | 25.3 |
| Water content, *w* (%) | 23.5 |

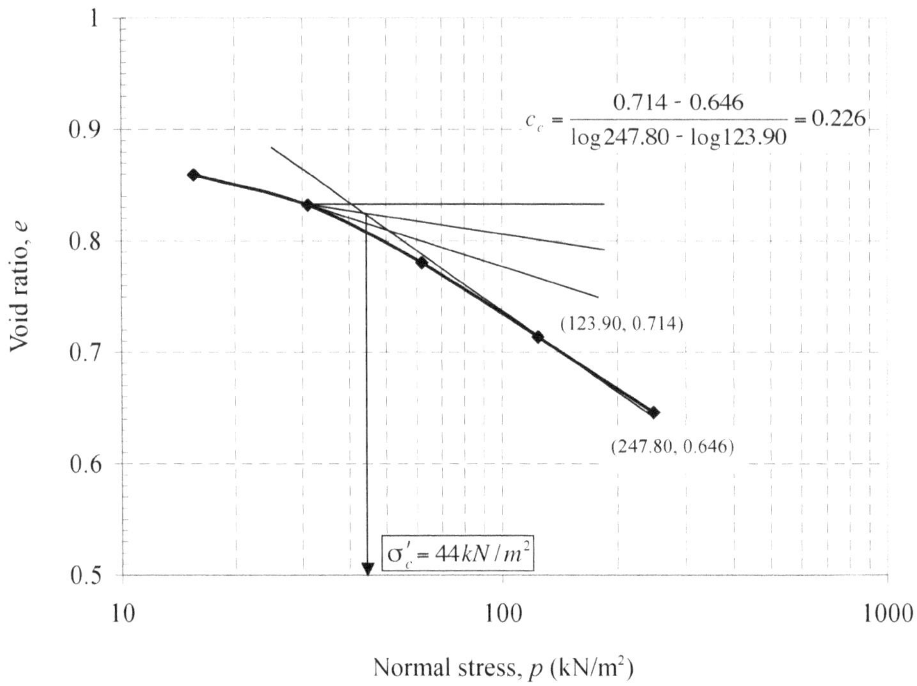


Figure 10-4. Consolidation e-logp curve.

Using the Casagrande Method, the preconsolidation pressure is 44kN/m$^2$; the compression index is 0.226.

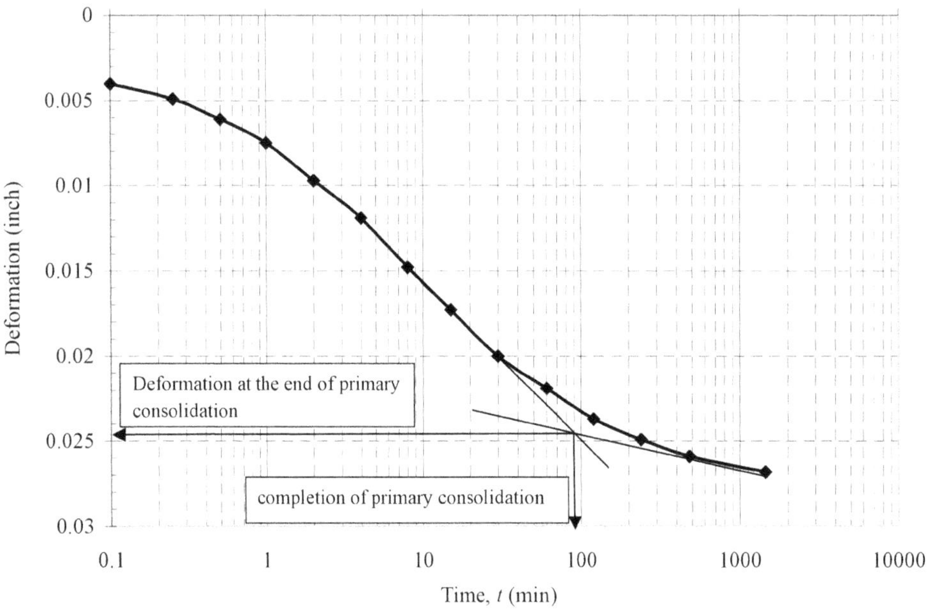


Figure 10-5. Time-deformation curve.

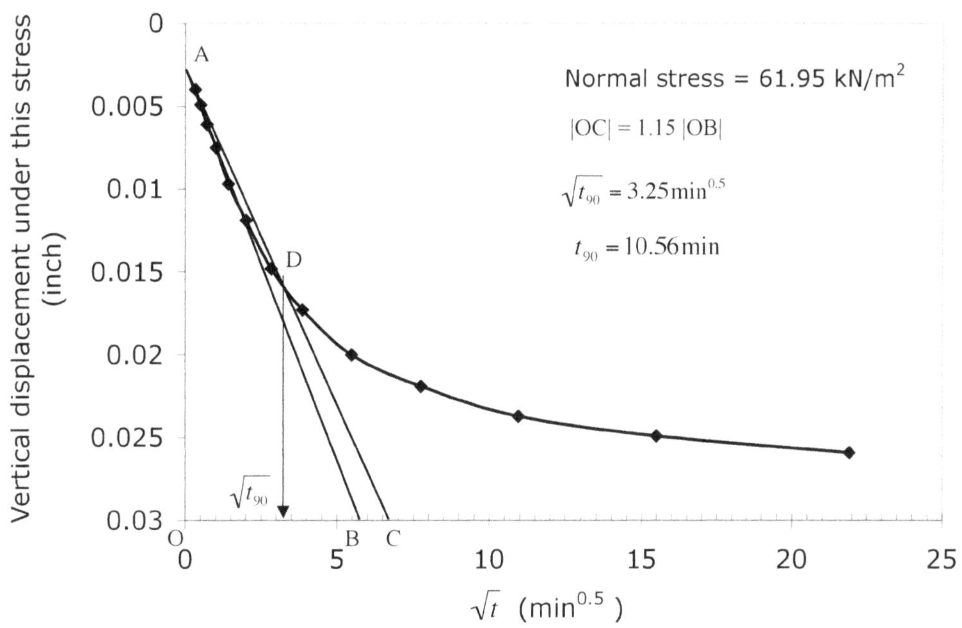


Figure 10-6. Square-root-of-time method to obtain coefficient of consolidation, $c_v$, (using data in Table 10-6).

(1) Extend the straight-line portion of the curve and draw line AB that intercepts the abscissa at point B.

(2) From the same origin A, draw straight line AC that intercepts the abscissa at point C, making |OC|=1.15|OB|.

(3) AC intercepts the curve at point D, which corresponds to the time of 90% of consolidation. The time factor, $T_v$, for 90% of consolidation is 0.848.

(4) Given:

$$T_{v(90)} = \frac{c_v t_{90}}{H_{dr}^2}$$

So:

$$c_v = \frac{T_{v(90)} H_{dr}^2}{t_{90}}$$

where:

$H_{dr}$ = the maximum drainage distance, for the clay specimen with porous stones at the top and the bottom,

$H_{dr} = \frac{1}{2} H_c$, and

$H_o$ = specimen thickness.

From Figure 10-6, $t_{90}$ = 10.56 minutes; $H_o$ = 1.0 inch = 2.54 cm, so

$$c_v = \frac{T_{v(90)} H_{dr}^2}{t_{90}} = \frac{0.848 \times (\frac{1}{2} \times 2.54cm)^2}{10.56\min} = 0.130 cm^2/\min$$

Name ______________________________

# Test 10: Consolidation Test Using Incremental Loading

Question 1: What is consolidation?

Question 2: Use the Casagrande method, find the preconsolidation pressure in the following e-logp chart.

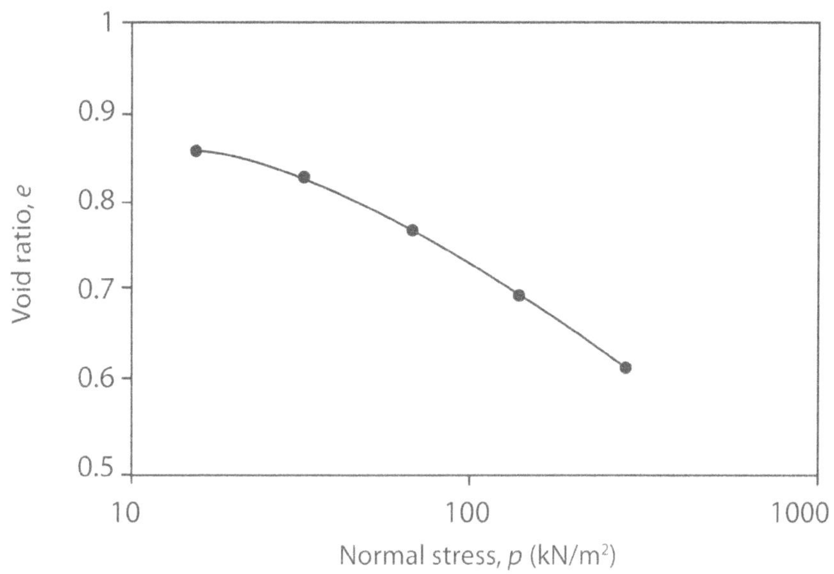

Question 3: Use the square-root-of-time method, find the coefficient of consolidation in the following consolidation graph.

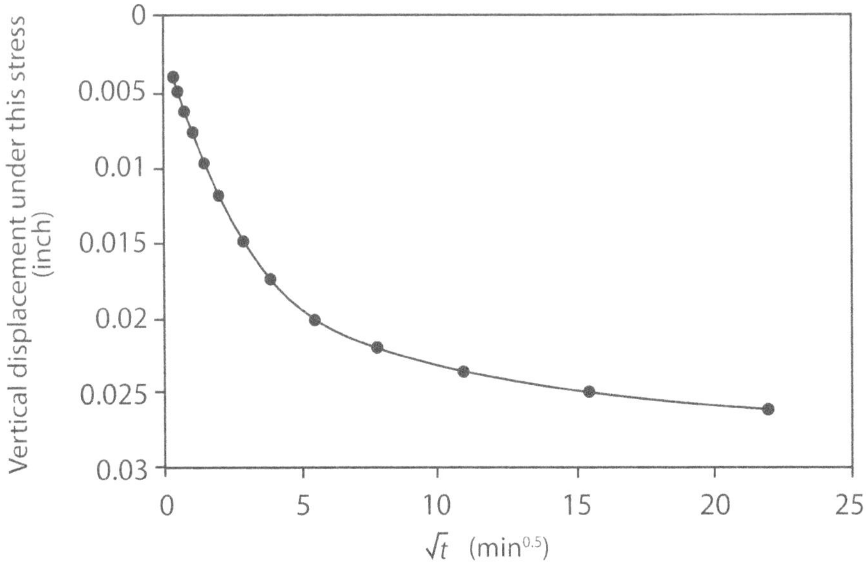

# TEST 11

# Direct Shear Test

## 1. Introduction

The objective of this test is to determine the shear strength of soils – the cohesion and the internal friction angle based on total stress.

The test method consists of placing a test specimen in the direct shear device, applying a predetermined normal stress, and displacing one frame horizontally with respect to the other at a constant rate of shear deformation. Consolidation and drainage during the shearing can be accommodated to obtain consolidated drained shear strength. Since the failure is forced to occur on a horizontal plane at the middle of the specimen, the obtained shear strength may not be the minimum shear strength of the specimen.

For each direct shear test under a certain normal stress, the shear strength, which is the maximum shear resistance, can be obtained. At least three tests are needed to define the Mohr-Coulomb failure envelope, which is assumed to be a straight line that best fits the test points, as shown in Figure 11-1. Each point represents the shear stress at failure under a certain normal stress. From the Mohr-Coulomb failure envelope, the cohesion (the intercept of the best-fit straight line with the ordinate) and the internal friction angle can be obtained.

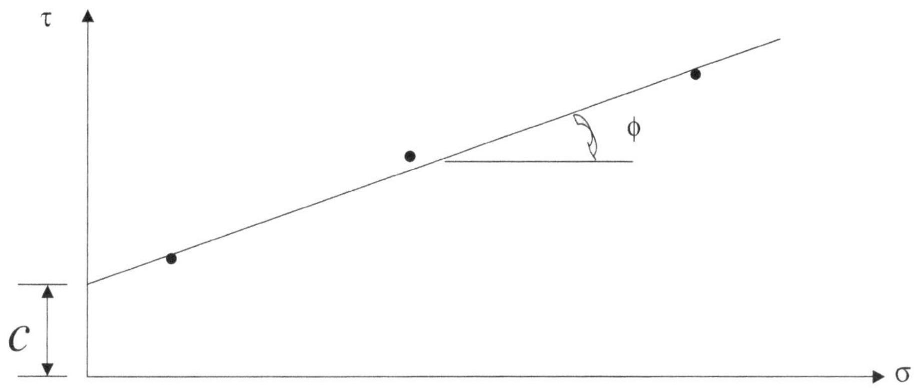


Figure 11-1. Mohr-Coulomb failure envelope obtained from direct shear tests.

## 2. References

- ASTM D 3080-04 "Standard Test Methods for Direct Shear Test of Soils Under Consolidated Drained Conditions," *2006 Annual Book of ASTM Standards*, Volume 04.08, Soil and Rock (I). ASTM International.

## 3. Soils and Other Materials

- The test can be conducted on all types of soils that can be undisturbed, remolded, or compacted.
- Distilled or demineralized water (if consolidation is conducted)

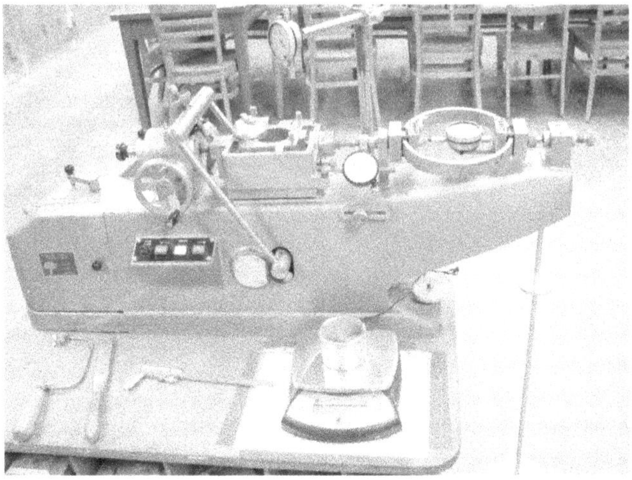

Figure 11-2. Direct shear apparatus (strain-controlled).

## 4. Apparatus

(1) Direct shear device, which includes the following:
- Shear box (circular or square) that is divided vertically into two halves of equal thickness which are fitted with alignment screws and gap screws; the gap screws control the gap between the top and bottom halves
- Shear device that holds the specimen and applies shear force to the specimen; it should provide the rate of displacement from 0.0001 to 0.04 in/min (0.0025 to 1.0 mm/min)
- Loading device that applies normal force through dead weight or pneumatic loading device
- Vertical deformation indicator: a dial gauge or an LVDT transducer with readout
- Horizontal deformation indicator: a dial gauge or an LVDT transducer with readout
- Shear force measurement: a proving ring with dial gauge or a load cell with readout
- Porous stones, if consolidation is to be performed before and during the shear

(2) Equipment for remolding or compacting specimen and trimming device (spatula, straightedge, wire saw)

(3) Water content measurement: drying oven, thermometer, water content container

(4) Miscellaneous apparatus: balance readable to the 0.1% of the total mass of the specimen, timer, and caliper.

## 5. Preparation of Test Specimen

- The specimen dimension requirements: (1) minimum specimen diameter or width: 2.0 inches (50 mm) and not less than 10 times the maximum particle size diameter; (2) minimum initial specimen thickness: 0.5 inches (12 mm) and not less than 6 times the maximum particle diameter; (3) minimum specimen diameter-to-thickness or width-to-thickness ratio: 2:1.

- If using an undisturbed specimen: trim the specimen to fit the shear box and to minimize disturbance and loss of moisture.
- If using a compacted specimen: the specimen can be compacted in layers inside the shear box using appropriate tools. Prior to compaction, the soil should be well mixed with a specified amount of water and cured according to the following guide:

| Soil Classification | Minimum Curing Time (h) |
|---|---|
| SW, SP | No requirement |
| M | 3 |
| SC, ML, CL | 18 |
| MH, CH | 36 |

## 6. Consolidometer Calibration

If the consolidation is required before shearing, the measured vertical deformation in the test must be corrected due to the compressibility of the porous disk, the filter paper, and the consolidometer.

(1) Select a copper or hard-steel disk of the same height as the soil specimen and a slightly smaller diameter than the shear box. Moisten the porous stone and the filter paper (if used).
(2) Assemble the shear box with the porous stones, the filter paper, and the disk.
(3) Load and unload the shear box, as will be performed in the consolidation test. Wait for 2 minutes and measure the deformation for each load applied.
(4) Plot or tabulate the corrections at each load applied.

## 7. Test Procedure

(1) Measure the inside dimension of the shear box: diameter and height, which are the dimensions of the soil specimen. The height should exclude the thickness of any (porous) plates used.
(2) Assemble the shear box:
    - Assemble the two halves of the shear box using the alignment screws and the gap screws. The gap screws control the gap between the top and the bottom halves. If consolidation is performed, do not open the gap until the consolidation is completed; if consolidation is not required, adjust the gap screws to form a gap between the shear box halves to approximately 0.025 inches (0.64 mm).

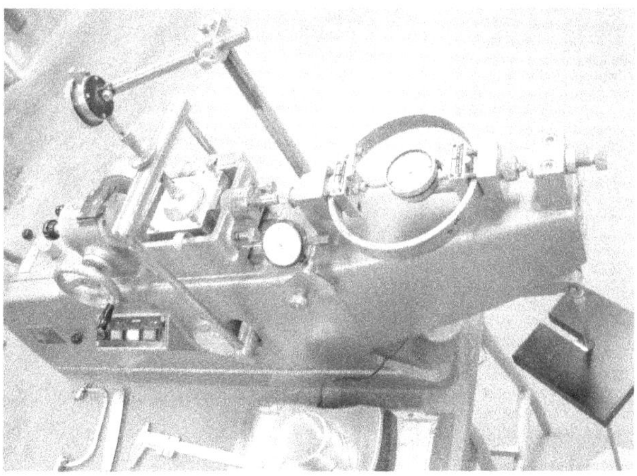

Figure 11-3. Shear test in progress.

- A light coating of grease applied between the halves may be used to reduce the friction during shear.

(3) Place the specimen in the shear box.
- If consolidation is to be performed before shear, wet porous plates should be used at the bottom and the top of the specimen.
- Measure the mass of the empty shear box.
- If using an undisturbed specimen, carefully slide the specimen in the shear box, using a minimum of force.
- If using a compacted specimen, compact the soil in layers (e.g., 3 layers) using approximately an equal amount of soil per layer. The layer boundary should not coincide with the shear plane defined by the shear box halves, unless this is the stated purpose for a particular test.
- If a porous plate is used at the top of the sample, the plate should be flush with the top of the shear box.
- Measure the mass of the shear box with the soil specimen.

(4) Place the assembled shear box in the shear box bowl, which is a metallic box that supports the shear box. Secure the shear box using the pins on the shear box bowl.

(5) Place the loading cap on the shear box, and place the vertical force loading yoke into position.

(6) Apply a small normal seating pressure (e.g., 5kPa or 100 lbf/ft$^2$) to the specimen. Verify that all components of the loading system are seated and aligned.

(7) Attach and adjust the vertical displacement measurement, the horizontal displacement measurement, and the proving ring or load cell to zero.

(8) (Optional Step). If consolidation is required, fill the shear box bowl with water and keep it full during the entire test. Apply a specified normal load (or increments thereof) and record the vertical displacement dial reading with time.

(9) Remove the alignment screws. If consolidation is performed, at the end of the consolidation, adjust the gap screws to form a gap between the shear box halves to approximately 0.025 inches (0.64 mm). Then remove the gap screws. **It is important to take out all the screws before shear.**

(10) Shear the specimen at an appropriate displacement rate. For drained shear, a slow rate should be used so that no excess pore pressure develops during shear.
- The displacement rate can be achieved by adjusting the gear wheel, gear lever, or motor speed.
- For clean dense sands in a drained test, 10 minutes may be used as the shear duration.
- For dense sands with more than 5% fines in drained test, 60 minutes may be used.
- For an undrained test, a quick shear can be performed, taking less than 15 minutes.

(11) During the shear, record the horizontal and vertical displacements and proving ring or load cell reading with time. After the proving ring or the load cell reading decreases or remains constant, it indicates shear failure. Take additional 3-5 readings to achieve a bell-shaped stress-displacement curve. Some soils may not display a well-defined bell shape.

(12) After specimen failure, stop shear. Remove the load and take out the shear box from the shear box bowl.
- For cohesive specimens, separate the shear box halves with a sliding motion along the failure plane. Photograph or sketch the failure surface.

(13) Measure the water content of the specimen.

(14) Repeat the test using a vertical load of two times the previous load. Three or more specimens shall be tested under different normal loads to define the Mohr-Coulomb failure envelope.

## 8. Data Collection and Calculation

Dimension of shear box:

| | |
|---|---|
| Inside diameter, | $D =$ |
| Cross-sectional area, | $A =$ |
| Height, | $H =$ |
| Soil sample volume, | $V =$ |

Assumed specific gravity of the sand: $G_s =$

Table 11-1. Dry Density and Void Ratio

| | **Test 1** | **Test 2** | **Test 3** |
|---|---|---|---|
| Mass shear box (g) | | | |
| Mass of shear box and specimen (g) | | | |
| Mass of specimen, $M$ (g) | | | |
| Moist density, $\rho = \frac{M}{V}$ (g/cm$^3$) | | | |
| Water content, $w$ (%) (from Table 11-2) | | | |
| Dry density, $\rho_d = \frac{\rho}{1+\frac{w(\%)}{100\%}}$ (g/cm$^3$) | | | |
| Void ratio, $e = \frac{G_s\rho_w}{\rho_d} - 1$ | | | |

Table 11-2. Water Content Measurement (at the end of the test)

| | **Test 1** | **Test 2** | **Test 3** |
|---|---|---|---|
| Mass of container, $M_1$ (g) | | | |
| Mass of container with moist soil specimen, $M_2$ (g) | | | |
| Mass of container with dry soil specimen, $M_3$ (g) | | | |
| Mass of water, $M_w = M_2 - M_3$ (g) | | | |
| Mass of dry soil, $M_s = M_3 - M_1$ (g) | | | |
| Moisture content, $w = \frac{M_w}{M_s} \times 100\%$ (%) | | | |

Table 11-3. Deformation and Stress

Mass of dead weight, =
Load on the specimen, $N =$
Normal stress, $\sigma = \frac{N}{A} =$
Time of shear, $t =$
Total horizontal deformation, $\Delta L =$
Strain rate $= \frac{\Delta L}{t} =$

| Horizontal Displacement | Vertical Displacement | Proving Ring Dial Gauge or Load Cell Reading | Shear Force, $T$ (Converted from Proving Ring, if used) | Shear Stress, $\tau = \frac{T}{A}$ |
|---|---|---|---|---|
| | | | | |
| | | | | |
| | | | | |
| | | | | |
| | | | | |
| | | | | |
| | | | | |
| | | | | |
| | | | | |
| | | | | |
| | | | | |
| | | | | |

Table 11-4. Shear Test Summary

| Normal Stress, $\sigma$ | Maximum Shear Stress, $\tau_f$ |
|---|---|
| | |
| | |
| | |

## 9. Results and Analyses

(1) For each test, plot the graph of horizontal displacement vs. shear stress ($\tau$). Find the shear strength of the soil, which is the maximum shear stress.

(2) If required, plot the graph of horizontal displacement vs. vertical displacement, using the data from Table 11-3. The graph shows the possible sample compression and dilation during shear.

(3) Plot the Mohr-Coulomb failure envelope using the data from Table 11-4. Determine the angle of internal friction and the cohesion of the specimen.

# SAMPLE DATA COLLECTION, RESULTS, AND ANALYSES

Tested specimen is air-dried sandy soil.
Unconsolidated shear test is performed.

Dimension of shear box:

| | |
|---|---|
| Inside diameter, | $D$ = 5.934cm = 2.336in |
| Cross-sectional area, | $A$ = 27.656cm$^2$ = 4.286in$^2$ |
| Height, | $H$ = 4.482cm = 1.764in |
| Soil sample volume, | $V$ = 123.954cm$^3$ = 7.560in$^3$ |

Assumed specific gravity of the sand: $G_s$ = 2.65

Table 11-1. Dry Density and Void Ratio

| | Test 1 | Test 2 | Test 3 |
|---|---|---|---|
| Mass shear box (g) | 2566.1 | 2566.1 | 2566.1 |
| Mass of shear box and specimen (g) | 2777.9 | 2778.5 | 2777.4 |
| Mass of specimen, $M$ (g) | 211.8 | 212.4 | 211.3 |
| Moist density, $\rho = \dfrac{M}{V}$ (g/cm$^3$) | 1.71 | 1.71 | 1.7 |
| Water content, $w$ (%) (from Table 11-2) | 1.3 | 1.3 | 1.4 |
| Dry density, $\rho_d = \dfrac{\rho}{1+\dfrac{w(\%)}{100\%}}$ (g/cm$^3$) | 1.69 | 1.69 | 1.68 |
| Void ratio, $e = \dfrac{G_s\rho_w}{\rho_d} - 1$ | 0.57 | 0.57 | 0.58 |

Table 11-2. Water Content Measurement (at the end of the test)

| | Test 1 | Test 2 | Test 3 |
|---|---|---|---|
| Mass of container, $M_1$ (g) | 27.3 | 27.1 | 27.1 |
| Mass of container with moist soil specimen, $M_2$ (g) | 135.2 | 140.5 | 138.7 |
| Mass of container with dry soil specimen, $M_3$ (g) | 133.8 | 139.0 | 137.2 |
| Mass of water, $M_w = M_2 - M_3$ (g) | 1.4 | 1.5 | 1.5 |
| Mass of dry soil, $M_s = M_3 - M_1$ (g) | 106.5 | 111.9 | 110.1 |
| Moisture content, $w = \dfrac{M_w}{M_s} \times 100\%$ (%) | 1.3 | 1.3 | 1.4 |

Table 11-3. Deformation and Stress

Mass of dead weight, = 4kg

Load on the specimen, $Q = 5\times4\times9.81=196.2\text{N}$

Normal stress, $\sigma = \dfrac{Q}{A} = 70.9\text{kN/m}^2 = 10.3\text{psi}$

Time of shear, $t$ = 5min 48sec = 5.8min

Total horizontal deformation, $\Delta L$ = 7.4mm

Shear rate, $= \dfrac{\Delta L}{t} = 1.27\text{mm/min}$

| Horizontal Displacement (mm) | Vertical Displacement (mm) | Proving Ring Dial Gauge Reading, $L$ (0.002 mm) | Shear Force, $T$ =199.71L + 2.5562 (lb) (L in mm) | Shear Stress, $\tau = \dfrac{T}{A}$ (lb/in$^2$) |
|---|---|---|---|---|
| 0 | 0 | 0 | 2.56 | 0.60 |
| 0.5 | -0.022 | 21 | 10.94 | 2.55 |
| 1.0 | -0.032 | 22 | 11.34 | 2.65 |
| 1.5 | -0.038 | 23 | 11.74 | 2.74 |
| 2.0 | -0.040 | 23 | 11.74 | 2.74 |
| 2.5 | -0.040 | 23 | 11.74 | 2.74 |
| 3.0 | -0.040 | 23 | 11.74 | 2.74 |
| 3.6 | -0.038 | 30 | 14.54 | 3.39 |
| 4.1 | -0.050 | 50 | 22.53 | 5.26 |
| 4.6 | -0.104 | 91 | 38.90 | 9.08 |
| 5.1 | -0.100 | 129 | 54.08 | 12.62 |
| 5.6 | -0.040 | 151 | 62.87 | 14.67 |
| 5.8 | 0.006 | 158 | 65.66 | 15.32 |
| 6.1 | 0.056 | 162 | 67.26 | 15.69 |
| 6.6 | 0.160 | 164 | 68.06 | 15.88 |
| 7.1 | 0.270 | 156 | 64.86 | 15.13 |
| 7.4 | 0.300 | 144 | 60.07 | 14.02 |

Table 11-4. Deformation and Stress

Mass of dead weight, = 8kg

Load on the specimen, $Q = 5\times8\times9.81=392.4N$

Normal stress, $\sigma = \frac{Q}{A} = 141.8kN/m^2 = 20.6psi$

Time of shear, $t$ = 6min 24sec = 6.4min

Total horizontal deformation, $\Delta L$ = 8.1mm

Shear rate, $= \frac{\Delta L}{t} = 1.27mm/min$

| **Horizontal Displacement (mm)** | **Vertical Displacement (mm)** | **Proving Ring Dial Gauge Reading, *L* (0.002 mm)** | **Shear Force, *T* =199.71L + 2.5562 (lb) (L in mm)** | **Shear Stress, $\tau = \frac{T}{A}$ (lb/in$^2$)** |
|---|---|---|---|---|
| 0 | 0 | 0 | 2.56 | 0.60 |
| 0.5 | -0.004 | 5 | 4.55 | 1.06 |
| 1.0 | -0.002 | 6 | 4.95 | 1.15 |
| 1.5 | -0.001 | 6 | 4.95 | 1.15 |
| 2.0 | 0.002 | 6 | 4.95 | 1.15 |
| 2.5 | 0.008 | 5 | 4.55 | 1.06 |
| 3.0 | 0.018 | 5 | 4.55 | 1.06 |
| 3.6 | 0.029 | 18 | 9.75 | 2.27 |
| 4.1 | 0.028 | 77 | 33.31 | 7.77 |
| 4.3 | 0 | 120 | 50.48 | 11.78 |
| 4.6 | -0.02 | 153 | 63.66 | 14.85 |
| 5.1 | -0.022 | 199 | 82.04 | 19.14 |
| 5.6 | 0.008 | 230 | 94.42 | 22.03 |
| 6.1 | 0.070 | 254 | 104.00 | 24.27 |
| 6.6 | 0.146 | 273 | 111.59 | 26.04 |
| 7.1 | 0.220 | 270 | 110.39 | 25.76 |
| 7.6 | 0.314 | 262 | 107.2 | 25.01 |
| 8.1 | 0.360 | 243 | 99.61 | 23.24 |

Table 11-5. Deformation and Stress

Mass of dead weight, = 12kg

Load on the specimen, $Q$ = 5×12×9.81=588.6N

Normal stress, $\sigma = \frac{Q}{A} = 221.7 \text{kN/m}^2 = 30.9 \text{psi}$

Time of shear, $t$ = 7min 12sec = 7.2min

Total horizontal deformation, $\Delta L$ = 9.1mm

Shear rate, $= \frac{\Delta L}{t} = 1.27 \text{mm/min}$

| **Horizontal Displacement (mm)** | **Vertical Displacement (mm)** | **Proving Ring Dial Gauge Reading, $L$ (0.002 mm)** | **Shear Force, $T$ =199.71L + 2.5562 (lb) (L in mm)** | **Shear Stress, $\tau = \frac{T}{A}$ (lb/in$^2$)** |
|---|---|---|---|---|
| 0 | 0 | 0 | 2.56 | 0.60 |
| 0.5 | -0.016 | 5 | 4.55 | 1.06 |
| 1.0 | -0.024 | 6 | 4.95 | 1.15 |
| 1.5 | -0.034 | 7 | 5.35 | 1.25 |
| 2.0 | -0.038 | 7 | 5.35 | 1.25 |
| 2.5 | -0.042 | 6 | 4.95 | 1.15 |
| 3.0 | -0.044 | 6 | 4.95 | 1.15 |
| 3.6 | -0.052 | 66 | 28.92 | 6.75 |
| 4.1 | -0.122 | 141 | 58.87 | 13.74 |
| 4.6 | -0.160 | 169 | 70.05 | 16.34 |
| 5.1 | -0.170 | 146 | 60.87 | 14.20 |
| 5.6 | -0.178 | 196 | 80.84 | 18.86 |
| 6.1 | -0.196 | 260 | 106.40 | 24.83 |
| 6.6 | -0.192 | 310 | 126.37 | 29.48 |
| 7.1 | -0.162 | 346 | 140.75 | 32.84 |
| 7.6 | -0.112 | 367 | 149.14 | 34.80 |
| 8.1 | -0.040 | 367 | 149.14 | 34.80 |
| 8.6 | 0.014 | 357 | 145.14 | 33.86 |
| 9.1 | 0.038 | 342 | 139.15 | 32.47 |

(1) For each test, plot the graph of horizontal displacement vs. shear stress ($\tau$). Find the shear strength of the soil, which is the maximum shear stress.

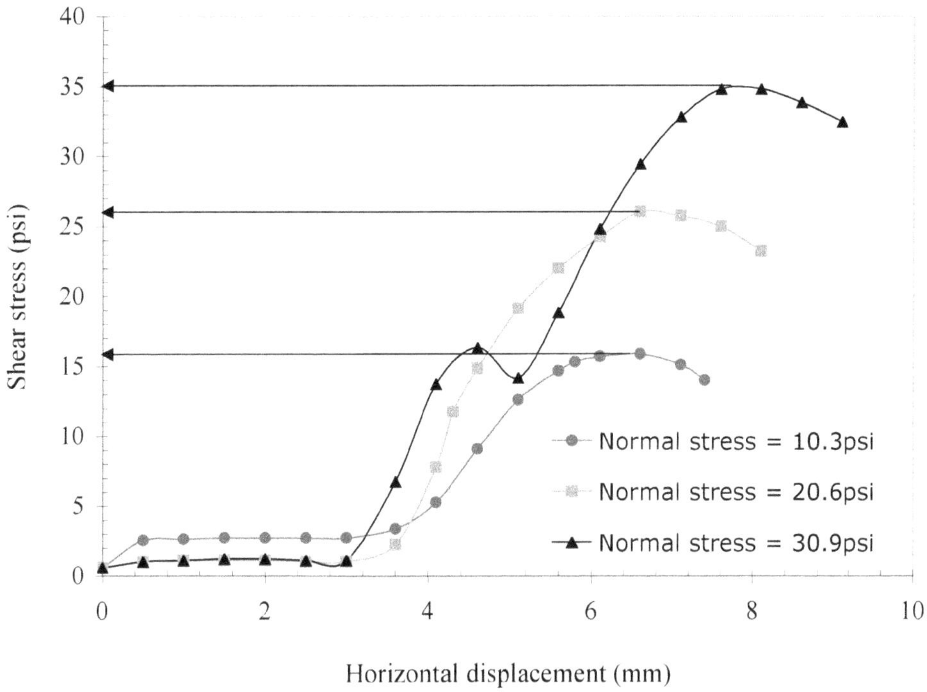


Figure 11-4. Shear stress development with horizontal displacement.

Table 11-6. Shear Test Summary

| **Normal Stress, $\sigma$ (lb/in$^2$)** | **Maximum Shear Stress, $\tau_f$ (lb/in$^2$)** |
|---|---|
| 10.3 | 16.0 |
| 20.6 | 26.0 |
| 30.9 | 35.0 |

(12) If required, plot the graph of horizontal displacement vs. vertical displacement, using the data from Tables 11-3 to 11-5. The graph shows the possible sample compression and dilation during shear.

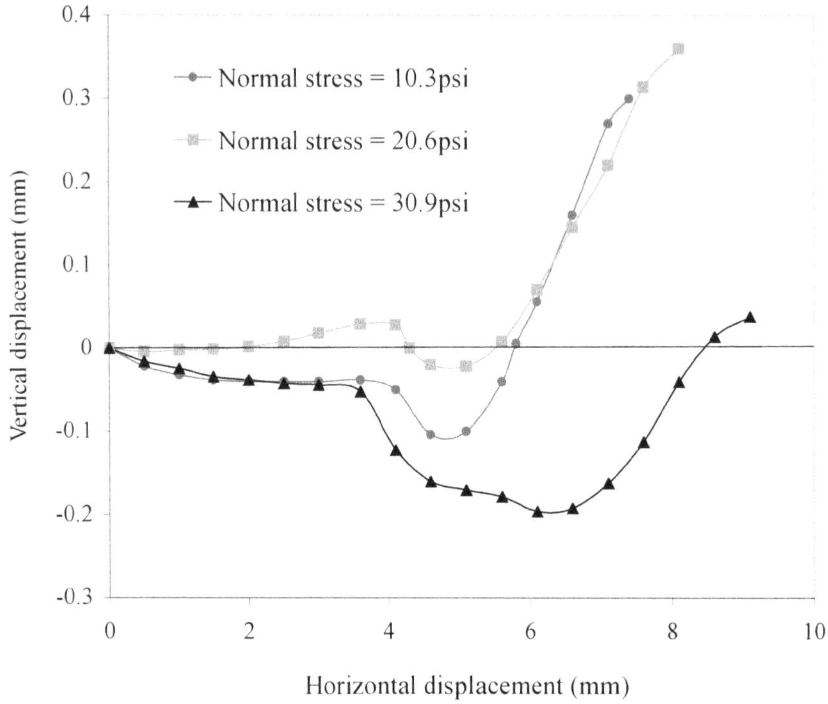


Figure 11-5. Specimen vertical displacement during shearing.

(3) Plot the Mohr-Coulomb failure envelope using the data from Table 11-6. Determine the angle of internal friction and the cohesion of the soil.

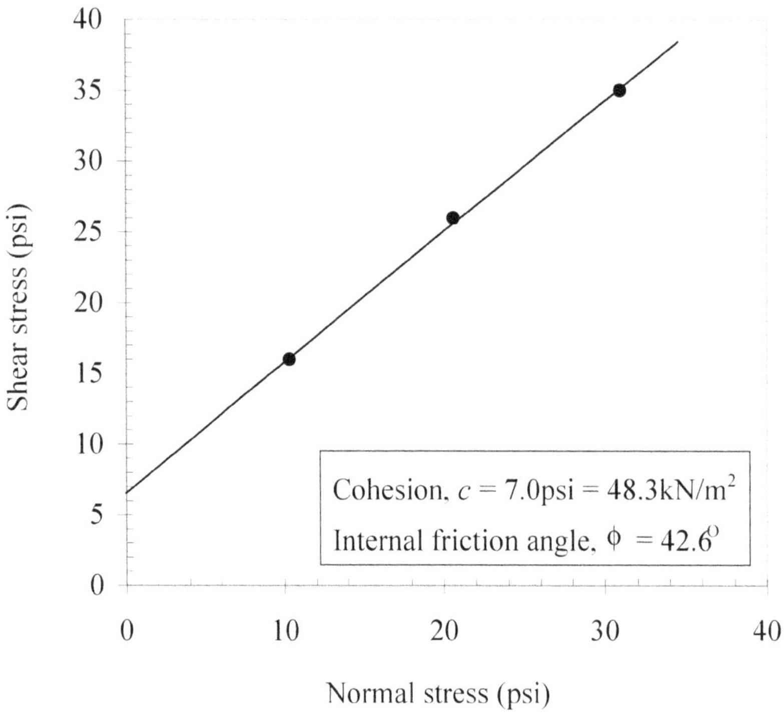


Figure 11-6. Mohr-Coulomb failure envelope.

**Cohesion, $c$ = 6.5 psi = 44.8 kN/m$^2$**
**Internal friction angle, $\phi$ = 42.6°**

Name ______________________________

# Test 11: Direct Shear Test

Question 1: What is the purpose of conducting a direct shear test?

Question 2: List the Mohr-Coulomb failure criterion and define the terms in the expression.

Question 3: What is the approximate gap between the two halves of the shear box in a direct shear test?

Question 4: For a strain-controlled direct shear apparatus, what is the ASTM-specified range of the displacement rate of the shear box?

# TEST 12

# Unconfined Compression Test on Cohesive Soils

## 1. Introduction

The objective of this test is to determine the unconfined compressive strength of cohesive soils in undisturbed, remolded, or compacted condition using a strain-controlled axial load. In this test method, unconfined compressive strength is taken as the maximum load attained per unit area or the load per unit area at 15% axial strain, whichever occurs first during the test.

The unconfined compressive strength test is conceptually a special type of triaxial test with the confining pressure as zero. The test usually completes within 15 minutes, so it is considered an undrained condition. The Mohr's circle for the test result is shown in Figure 12-1. The Mohr-Coulomb failure envelope is horizontal and tangent to the circle. Therefore, the undrained internal friction angle is zero, and the undrained cohesion $c_u$ is half of the compressive stress at failure, $q_u$, $q_u$ is the undrained shear strength.

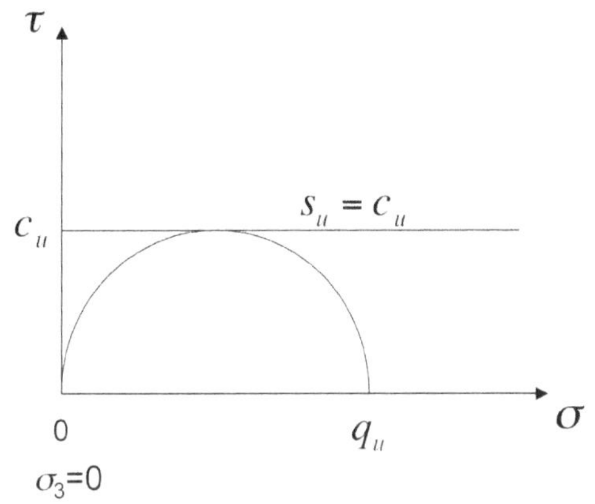


Figure 12-1. Mohr's circle of a specimen at failure in an unconfined compression test.

## 2. References

- ASTM D 2166-00 "Standard Test Methods for Unconfined Compressive Strength of Cohesive Soils," *2006 Annual Book of ASTM Standards*, Volume 04.08, Soil and Rock (I). ASTM International.

## 3. Soils and Other Materials

- Cohesive soils

## 4. Apparatus

- Compression device with deformation indicator, such as vertical displacement dial gauge or LVDT transducer; the compression device includes a motorized pedestal that can move up and down at a specified speed and a proving ring or load cell that records the load applied on the specimen
- Balance that can measure the mass of the specimen within 0.1% of its total mass
- Moisture content container with lid (2)
- Drying oven maintaining a constant temperature of 110±5°C
- Trimming device (trimming turntable, wire saw, straightedge)
- Timer, caliper

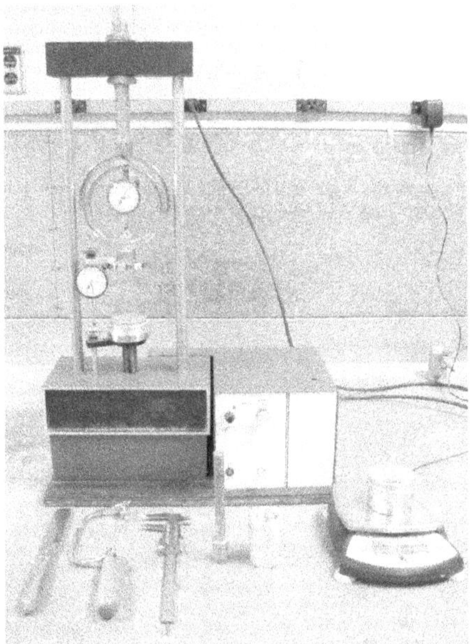

Figure 12-2. Unconfined compression test apparatus and tools.

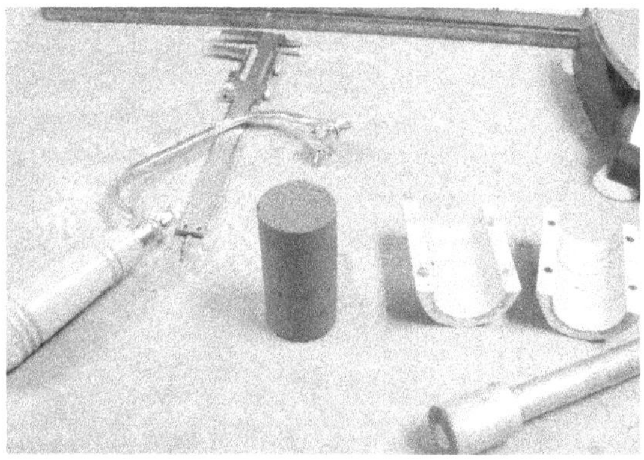

Figure 12-3. Remolded soil specimen.

## 5. Preparation of Test Specimen

- Specimen shall have a minimum diameter of 30 mm (1.3 inches). The height-to-diameter ratio shall be 2.0 to 2.5.
- The specimen can be from undisturbed, remolded, or compacted soils. To obtain the required diameter and height, the specimen can be trimmed using a trimming turntable, wire saw, or straightedge. The specimen shall have uniform circular cross section with ends perpendicular to the longitudinal axis of the specimen.

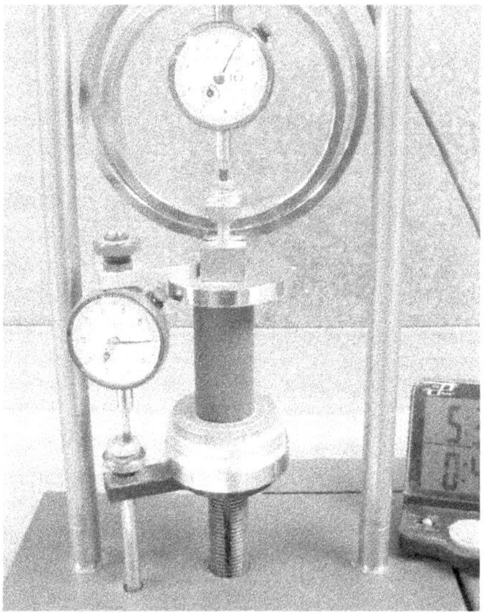

Figure 12-4. Specimen before compression.

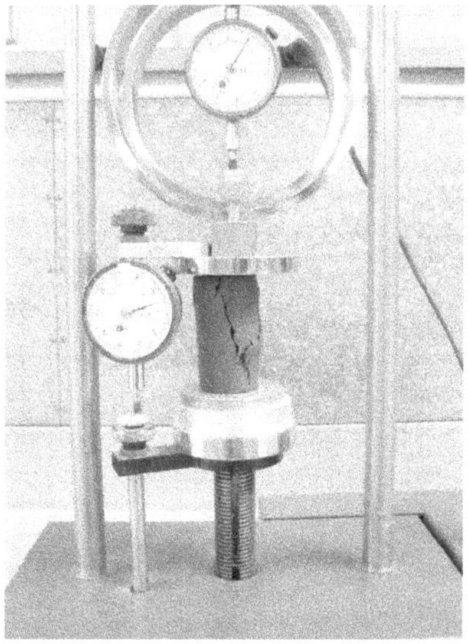

Figure 12-5. Specimen at failure.

## 6. Test Procedure

(1) Measure the mass of the specimen.

(2) Measure the dimensions of the specimen using a caliper:
- Measure the diameters at the lower, mid, and upper sections and calculate the average.
- Take at least three height measures (120° apart) and use the average value as the height.

(3) Place the specimen on the loading device and center the specimen on the pedestal. Adjust the loading device carefully so that the upper platen just makes contact with the specimen. Zero the deformation dial gauge or LVDT transducer (Figure 12-4).

(4) Turn on the loading device so that the pedestal moves up to compress the specimen. The axial strain shall be at a rate of 0.5~2.0%/min. Record the readings of the deformation dial, the proving ring dial or load cell, and the time. Continue loading until the load values decrease with increasing strain, or until 15% strain is reached.
- Note: The deformation and the load readings should be taken at the same time.
- Usually 10 to 15 readings are sufficient to define the shape of the stress-strain curve.
- The rate of strain should be chosen so that the time to failure does not exceed 15 minutes.
- If a proving ring is used, the constant that converts the proving ring dial reading to the load should be known (provided by the manufacturer).

(5) After the specimen fails, a failure surface can be visually seen on the specimen (Figure 12-5). Make a sketch or take a photo of the test specimen at failure, showing the slope angle of the failure surface.

(6) Carefully remove the specimen from the loading device and measure the water content of the specimen.

## 7. Data Collection and Calculation

Table 12-1. Measurement of Specimen's Initial Dimension

| | Measurement 1 | Measurement 2 | Measurement 3 | Average Value |
|---|---|---|---|---|
| Length, $L_0$ (cm) | | | | |
| Diameter, $D_0$ (cm) | | | | |

Initial cross-sectional area, $A_0$ =
Initial volume, $V_0$ =
Mass of specimen, =
Moist density, =

Proving ring constant (if used), $K$, to convert the proving ring dial reading to the load:
$K$ =

Table 12-2. Stress-strain of Specimen During Compression

Compression duration =

| Vertical Displacement Dial Reading, $\Delta L$ | Proving Ring or Load Cell Reading, $R$ | Axial Load, N | Total Strain, $\varepsilon$ (%) | Corrected Area, $A = \frac{A_0}{1-\varepsilon}$ | Stress, $\sigma = \frac{N}{A}$ |
|---|---|---|---|---|---|
| | | | | | |
| | | | | | |
| | | | | | |
| | | | | | |
| | | | | | |
| | | | | | |
| | | | | | |
| | | | | | |
| | | | | | |
| | | | | | |
| | | | | | |
| | | | | | |
| | | | | | |
| | | | | | |

Table 12-3. Final Moisture Content of the Specimen

| | |
|---|---|
| Mass of moisture can (g) | |
| Mass of moisture can and specimen (g) | |
| Mass of moisture can and specimen soil after oven-drying (g) | |
| Dry soil mass (g) | |
| Water in the soil (g) | |
| Water content, *w* (%) | |

## 8. Results and Analyses

(1) Complete Table 12-2 using the following approach:

(a) Calculate the axial strain, $\varepsilon$, to the nearest 0.1%, for each applied load:

$$\varepsilon = \frac{\Delta L}{L_0} \times 100\%$$

where:

$\Delta L$ = is the length change indicated by the vertical displacement dial reading.

(b) Calculate the average cross-sectional area (corrected area), $A$, at each applied load:

$$A = \frac{A_0}{1 - \varepsilon}$$

(c) Calculate the axial load, $N$, from the load cell or proving ring reading:

$$N = f(K, R)$$

where:

$K$ is the constant to convert the proving ring dial reading, $R$, to the axial load, $N$. The function is usually provided by the loading device manufacturer.

(d) Calculate the compressive stress, $\sigma$:

$$\sigma = \frac{N}{A}$$

(2) Plot the strain-stress graph and determine the maximum value of compressive stress or the compressive stress at 15% axial strain, whichever is secured first.

(3) Report the unconfined compressive strength, the strain at failure, and the undrained cohesion.

(4) Include a sketch or a photo showing the angle of the failure surface on the specimen.

## SAMPLE DATA COLLECTION, RESULTS, AND ANALYSES

Table 12-1. Measurement of Specimen's Initial Dimension

| | Measurement 1 | Measurement 2 | Measurement 3 | Average Value |
|---|---|---|---|---|
| Length, $L_0$ (cm) | 8.144 | 8.150 | 8.138 | 8.144 |
| Diameter, $D_0$ (cm) | 3.470 | 3.472 | 3.482 | 3.475 |

Initial cross-sectional area, $A_0$ = 9.484cm$^2$

Initial volume, $V_0$ = 77.239cm$^3$

Mass of specimen, = 135.5g

Moist density, = 1.75g/cm$^3$ = 109.5 lb/ft$^3$

In this test, a proving ring is used. The following relation is used to convert the proving ring dial gauge reading, $R$, to the axial load, $N$. The unit of $R$ is 0.0001;

$$N(lb) = 14330R$$

Table 12-2. Stress-strain of Specimen During Compression

| Vertical Displacement Dial Reading, $\Delta L$ (0.001 in) | Proving Ring or Load Cell Reading, $R$ (0.0001) | Axial Load, N (lb) | Total Strain, $\varepsilon$ (%) | Corrected Area, $A = \frac{A_0}{1-\varepsilon}$ (cm$^2$) | Stress, $\sigma = \frac{N}{A}$ (kN/m$^2$) | Stress, $\sigma = \frac{N}{A}$ (psi) |
|---|---|---|---|---|---|---|
| 10 | 2.1 | 3.01 | 0.31 | 9.51 | 14.08 | 2.04 |
| 20 | 4.2 | 6.02 | 0.62 | 9.54 | 28.08 | 4.07 |
| 30 | 6.8 | 9.74 | 0.94 | 9.57 | 45.29 | 6.57 |
| 40 | 9.5 | 13.61 | 1.25 | 9.60 | 63.09 | 9.15 |
| 50 | 12.6 | 18.06 | 1.56 | 9.63 | 83.45 | 12.10 |
| 60 | 16.5 | 23.64 | 1.87 | 9.66 | 108.90 | 15.79 |
| 70 | 19.6 | 28.09 | 2.18 | 9.69 | 128.99 | 18.70 |
| 80 | 21.0 | 30.09 | 2.50 | 9.72 | 137.75 | 19.97 |
| 90 | 23.5 | 33.68 | 2.81 | 9.75 | 153.71 | 22.29 |
| 100 | 25.0 | 35.83 | 3.12 | 9.79 | 162.86 | 23.61 |
| 110 | 25.8 | 36.97 | 3.43 | 9.82 | 167.52 | 24.29 |
| 120 | 26.2 | 37.54 | 3.74 | 9.85 | 169.59 | 24.59 |
| 130 | 26.2 | 37.54 | 4.05 | 9.88 | 169.07 | 24.51 |
| 140 | 25.2 | 36.11 | 4.37 | 9.91 | 162.14 | 23.51 |
| 150 | 23.6 | 33.82 | 4.68 | 9.95 | 151.25 | 21.93 |
| 160 | 21.5 | 30.81 | 4.99 | 9.98 | 137.37 | 19.92 |
| 170 | 20.0 | 28.66 | 5.30 | 10.01 | 127.40 | 18.47 |

Compression duration: t = 4.5 minutes,
Maximum strain = 5.3%,
Strain ratio = 1.2%/min.

Table 12-3. Final Moisture Content of the Specimen

| | |
|---|---|
| Mass of moisture can (g) | 27.1 |
| Mass of moisture can and specimen (g) | 137.0 |
| Mass of moisture can and specimen soil after oven-drying (g) | 119.1 |
| Dry soil mass (g) | 91.9 |
| Water in the soil (g) | 17.9 |
| Water content, *w* (%) | 19.5 |

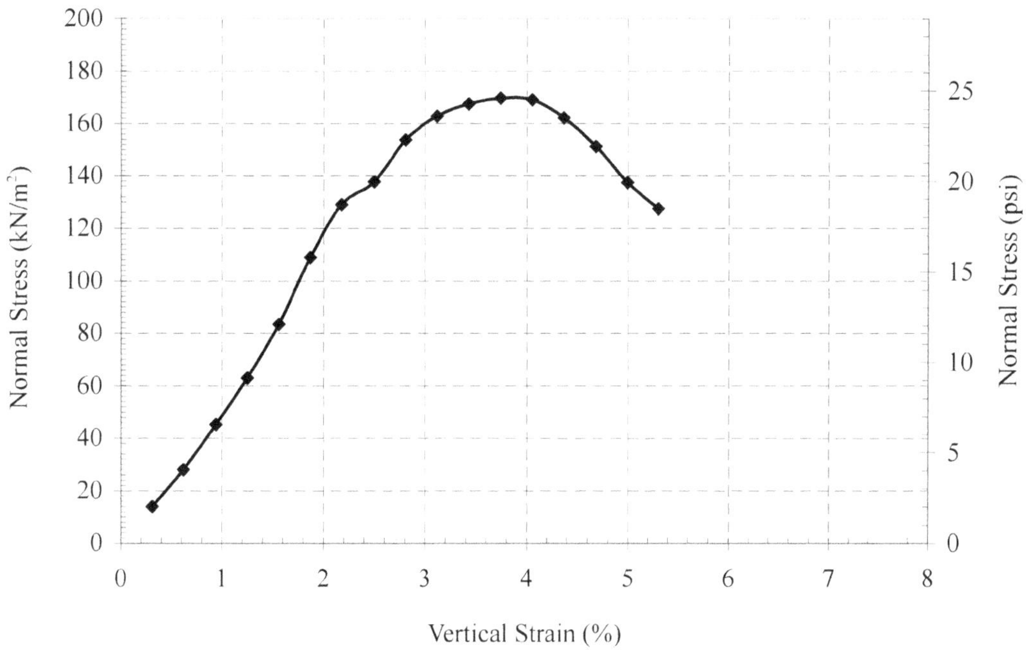


Figure 12-6. Stress-strain relationship during unconfined compression.

**The unconfined compressive strength,** $q_u = 169.6 kN/m^2 = 24.6 psi$

**The undrained cohesion,** $c_u = \frac{q_u}{2} = 84.8 kN/m^2 = 12.3 psi$

**Strain at failure,** $\varepsilon = 3.7\%$

Name ______________________________

# Test 12: Unconfined Compression Test on Cohesive Soils

Question 1: What are the key soil properties that the unconfined compression test can derive?

Question 2: What is the range of the axial compression rate that the unconfined compression test apparatus should provide?

Question 3: What is the relationship between the unconfined compressive strength and the undrained cohesion?

T E S T 13

# Permeability Test Using Flexible Wall Permeameter

## 1. Introduction

The objective of this test is to determine the coefficient of permeability (hydraulic conductivity) of water-saturated porous materials that have a hydraulic conductivity from $1\times10^{-4}$cm/s to $1\times10^{-9}$cm/s. The test is conducted using triaxial apparatus, which encases the specimen in a flexible membrane and applies confining pressure, and consolidation is permitted prior to the permeability measurement. The test minimizes side-wall leakage and simulates the in-situ stress condition of the specimen. The specimen can be undisturbed, remolded, or compacted.

There are six alternate methods or hydraulic systems that may be used to measure the permeability:

Method A – Constant head
Method B – Falling head, Constant tailwater elevation
Method C – Falling head, Rising tailwater elevation
Method D – Constant rate of flow (precision flow pump may be needed)
Method E – Constant volume - Constant head (by mercury)
Method F – Constant volume - Falling head (by mercury)

In this test illustration, Method A is used. The equipment setup is shown in Figure 13-1. The load frame, load transducer, and deformation transducer are not needed in the permeability measurement.

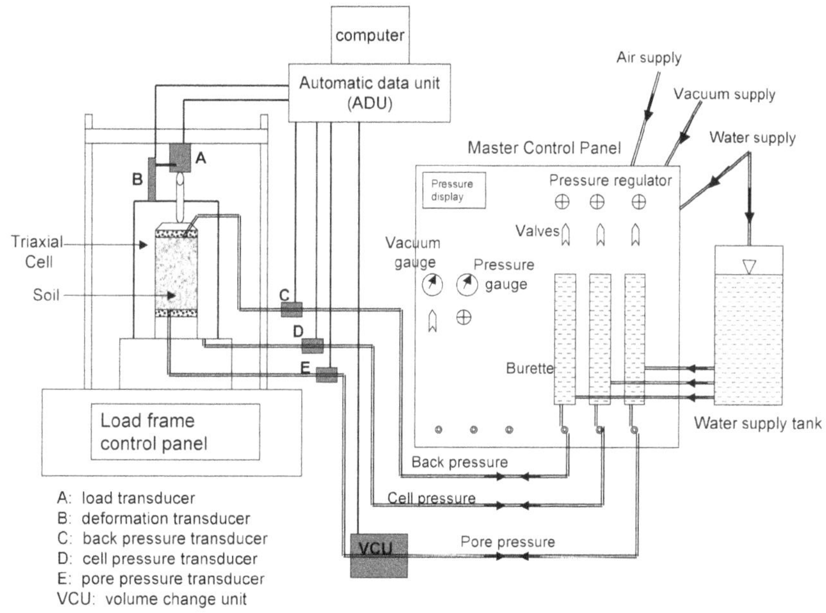


Figure 13-1. Triaxial equipment setup for permeability measurement.

The test method applies to one-dimensional, laminar flow in water-saturated porous media. The calculation of permeability is based on Darcy's law:

$$v = k \cdot i$$

where:

$v$ = flow velocity in soil,

$$v = \frac{q}{A}$$

where:

$q$ = flow rate (volume of water passing a unit cross-sectional area per unit of time), and
$A$ = cross-sectional area of soil specimen;

$$q = \frac{Q}{t}$$

where:

$Q$ = the volume of water passing through the soil specimen during time interval, $t$; and
$i$ = hydraulic gradient,

$$i = \frac{h}{L}$$

where:

$h$ = the total hydraulic head loss across the soil specimen with length, $L$.

Therefore, the coefficient of permeability can be calculated as follows:

$$k = \frac{QL}{Aht}$$

## 2. References

- ASTM D 5084-03 "Standard Test Methods for Measurement of Hydraulic Conductivity of Saturated Porous Materials Using a Flexible Wall Permeameter," *2006 Annual Book of ASTM Standards*, Volume 04.08, Soil and Rock (I). ASTM International.

## 3. Soils and Other Materials

- Fine-grained soils (silt or clay)
- Tap water

## 4. Apparatus

- Water supply, air supply, and vacuum supply sources
- A master control panel that can supply and regulate pressurized water, pressurized air, and vacuum; the panel should include burettes to measure the volume of influent to and effluent from the specimen
- A permeameter cell that encases the specimen; the cell should have a rigid and impermeable top cap and a base, a piston to hold the specimen in place, and outlets to the control panel
- Water supply tank that can sustain pressure up to 150 psi (1034kPa).

- Three water-pressure transducers and two volume-change transducers with data acquisition systems or readouts (volume-change transducer may not be needed if inflow and outflow can be measured using burettes)
- Flexible membrane whose diameter is 90–95% of that of the specimen, two porous stones whose diameter difference with that of the specimen is ±5%, filter paper, and four o-rings whose diameter is less than 90% of that of the specimen.
- Sample compaction equipment, such as tamper and mold of appropriate dimension (not needed if using an undisturbed specimen)
- Sample extruder (not needed if the sample compaction mold is split type)
- Trimming equipment (trimming turntable, wire saw, straightedge)
- Membrane tester (to test for tears or holes in the membrane)
- Membrane expander used to place the membrane onto the specimen
- O-ring placing tool
- Balance
- Water content container
- Drying oven
- Timer
- Caliper
- Thermometer
- Rigid tubings (so they do not expand under pressurized water)

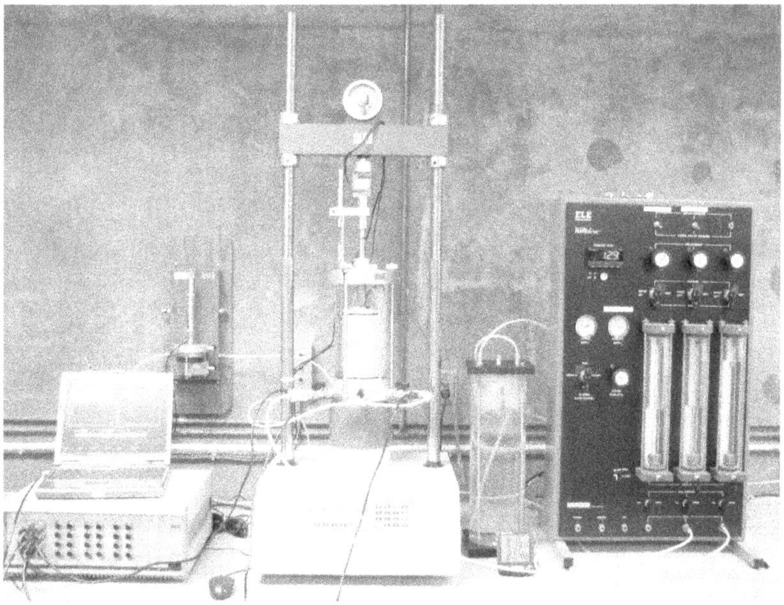

Figure 13-2. Flexible wall permeameter test setup.

## 5. Preparation of Test Specimen

The specimen shall have a minimum diameter of 25 mm (1.0 in) and a minimum height of 25 mm. The diameter and height of the specimen shall each be at least 6 times greater than the largest particle size in the specimen.

**If using undisturbed specimens:** Specimens obtained by tube sampling may be tested without trimming except for cutting the end surfaces.

**If using laboratory-compacted specimens:** The soil should be first moistened at the specified water content and cured for 16 hours if the soil is silt or clay. When curing, the soil should be covered to prevent moisture loss. The soil is then compacted in a mold (a split mold is preferred) using several layers (Figure 13-3). Each layer should have the same amount of soil and be compacted to the same thickness to achieve homogeneity. The surface of each previously compacted layer shall be lightly scarified with a straightedge to minimize the effect of strata.

The test specimen shall be mounted immediately in the permeameter after trimming to prevent moisture loss.

Figure 13-3. Specimen compaction device.

Figure 13-4. Mounting the specimen.

## 6. Test Procedure

(1) Mount the specimen (Figure 13-4):
- Prepare de-aired water in the water supply tank by applying vacuum to the tank for at least 30 minutes. The vacuum should be high enough to initially generate air bubbles in the tank.
- Place the membrane on the membrane tester and supply about 1 psi to inflate the membrane slightly to check for possible leaks.
- After trimming the specimen, measure the diameter, length, and mass of the specimen.
- In the bottom-up sequence, place a porous stone, filter paper, specimen, filter paper, porous stone, and top cap on the base plate.
- Place the membrane in the membrane expander and apply vacuum to expand the membrane to adhere to the inside wall of the expander. Then slide the membrane over the specimen, and release the vacuum.
- Place two O-rings to seal the membrane to the base and two O-rings to seal the membrane to the top cap, using an O-ring expander.
- Attach flow tubing(s) to the top cap. The flow tubing connects to the back-pressure port.
- Enclose the specimen with the clear cylinder and the cell top. Tighten the cell top. Bring the piston into contact with the specimen cap. The piston helps keep the specimen vertical. Record the piston location.

(2) Set up transducers and flow lines:
- Connect the cell-pressure transducer, pore-pressure transducer, and back-pressure transducer to the corresponding ports, which are connected to the control panel via rigid tubing. The transducers should have readouts, or the transducers can be con-

nected to data acquisition units so the readings can be displayed on a computer (appropriate software may be needed).

**Note:** The pore-pressure transducer should be between a valve and the permeameter cell.

- Connect the volume-change unit between the control panel and the back-pressure port, and connect a second volume-change unit between the control panel and the pore-pressure port.
- Fill the permeameter cell with water until all the air bubbles are expelled from the cell.
- Remove any air bubbles in the flow lines by applying a small pressure (1 psi) and opening the water-bleeding ports. The valves to the specimen should be closed during this process. Then reduce the pressure to zero.
- Reset the transducers to zero.
- Close all pressure valves.

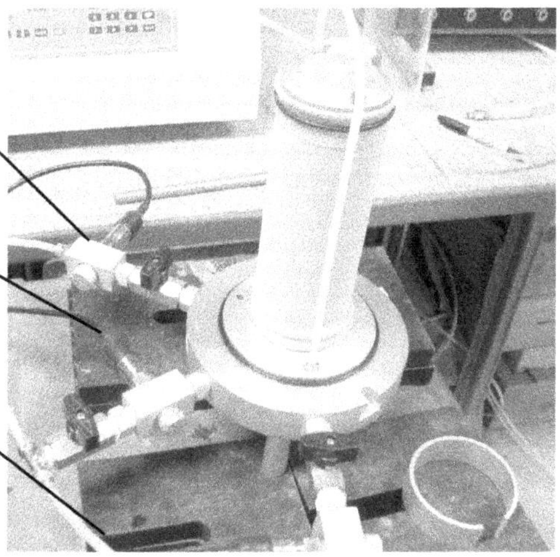

Figure 13-5. Test specimen in membrane.

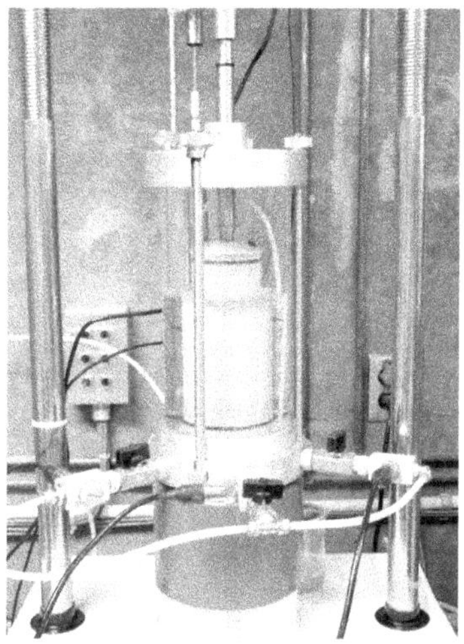
Figure 13-6. Filling the cell with water.

(3) Saturation:

When the initial degree of saturation is less than 85%, applying a low vacuum to the specimen may help reduce continuous air in the specimen. When using a vacuum, the top porous stone and filter paper should be dry. The steps are detailed below:

- Empty the burette that connects to the back-pressure valve, and adjust the vacuum in the burette to 35kPa (5 psi).
- Open the back-pressure valve and keep the pore- and cell-pressure valves closed. Vacuum is applied to the specimen from the top. Maintain the vacuum for 10 minutes; then reduce the vacuum to 20kPa (3 psi).
- Increase the cell pressure to a desired value, such as 70kPa (10 psi).
- Slowly open the cell-pressure valve and wait 2 minutes for stabilization.
- Slowly open the pore-pressure valve. Water will enter the specimen from the bottom under vacuum. If needed, a small pore pressure (e.g., 35kPa) can be applied at the specimen bottom to facilitate saturation; if so, the cell pressure should be at least 1 psi more than the pore pressure to prevent membrane bloating.
- When water appears in the burette that is connected to the top of the specimen, slowly reduce the vacuum to zero.
- Close both the pore- and back-pressure valves, and fill de-aired water in the burette that is connected to the back-pressure outlet.

- Adjust both the pore and back pressure. The pore and back pressures should be the same. The difference between the cell pressure and the pore/back pressure should be between 15 and 35kPa (2 and 5 psi).
- Open the pore- and back-pressure valves slowly and simultaneously. Water will enter into the specimen from the top and the bottom. The air will dissolve into the de-aired water. This process may take 24 hours to several days, depending on the soil.
- Check $B$ value to verify saturation:

$$B = \frac{\Delta u}{\Delta \sigma_3}$$

where:

$\Delta\sigma_3$ = cell pressure increase, and

$\Delta u$ = pore pressure increase.

First close the cell-, pore-, and back-pressure valves; then increase the cell pressure by 70kPa (10 psi), and open the cell-pressure valve. Wait for 2 minutes to stabilize. Read the pore-pressure reading and calculate the pore-pressure increase. The $B$ value can then be calculated. The specimen shall be considered saturated if $B \geq 0.95$. Otherwise, increase the pore and back pressure by 10 psi (same as $\Delta\sigma_3$); then open both the pore- and back-pressure valves. Wait for another 24 to 48 hours, and check the $B$ value again. Repeat this process until $B \geq 0.95$.

(4) Consolidation

- Bring the piston into contact with the specimen and record the piston location.
- Close the cell-, pore-, and back-pressure valves. Adjust the cell pressure to achieve the desired effective stress, which is the cell pressure minus the pore pressure.
- Open the cell-pressure valve and wait 2 minutes for the pressure to stabilize. Then open both the pore- and back-pressure valves to start consolidation. Water should slowly drain into both the pore- and back-pressure burettes.
- Wait 24 hours for consolidation to complete. At the end of the consolidation, bring the piston into contact with the specimen cap. Record the piston location and calculate the piston movement so that the specimen height at the end of the consolidation can be calculated.

(5) Permeation

- Close both the pore- and back-pressure valves. Increase the pore or back pressure to achieve the appropriate hydraulic gradient. The flow direction can be upward or downward. Do not decrease both the pore and back pressure because dissolved air may come out of water at reduced pressure. The pressure increase should not exceed 95% of the effective stress. Excessive gradients should be avoided because high seepage pressure may consolidate the soil. The recommended maximum hydraulic gradients are as follows:

| **Hydraulic Conductivity (cm/s)** | **Recommended Maximum Hydraulic Gradients** |
|---|---|
| $1\times10^{-3}$ to $1\times10^{-4}$ | 2 |
| $1\times10^{-4}$ to $1\times10^{-5}$ | 5 |
| $1\times10^{-5}$ to $1\times10^{-6}$ | 10 |
| $1\times10^{-6}$ to $1\times10^{-7}$ | 20 |
| less than $1\times10^{-7}$ | 30 |

A higher gradient than the above-given values may be used if the higher gradient can be shown not to change the hydraulic conductivity.

- Record the initial readings of the burettes that connect to the pore- and back-pressure outlets. Open the pore- and back-pressure valves at the same time, start the timer, and measure the inflow and outflow in time duration, $\Delta t$. After sufficient amount of effluent is collected, close the pore- and back-pressure valves and stop the timer. Calculate the hydraulic conductivity, using

$$k = \frac{Q \cdot L}{A \cdot \Delta h \cdot \Delta t}$$

where:

$Q$ = the average of the inflow and outflow,
$L$ = the specimen height at the end of the consolidation,
$\Delta h$ = the average of the head loss across the specimen, $(h_1 + h_2)/2$,
$h_1$ = head loss across the specimen at the beginning of the measurement,
$h_2$ = head loss across the specimen at the end of the measurement,
$A$ = the cross-sectional area; to simplify the calculation, $A$ uses the original cross-sectional area at the beginning of the test,
$\Delta t$ = time duration.

- Repeat the measurement four times under the same hydraulic gradient. Calculate the average of the hydraulic conductivity.
- Measure the temperature and correct the hydraulic conductivity to that at 20°C.

## 7. Data Collection and Calculation

Table 13-1. Initial Specimen Dimension and Density

| | |
|---|---|
| Initial water content (%) | |
| Specific gravity, $G_s$ | |
| Length, $L_0$ (cm) | |
| Diameter, $D$ (cm) | |
| Mass of specimen, $M$ (g) | |
| Cross-sectional area, $A$ (cm$^2$) | |
| Volume, $V$ (cm$^3$) | |
| Moist density, $\rho$ (g/cm$^3$) | |
| Dry density, $\rho_d$ (g/cm$^3$) | |
| Void ratio, $e = \frac{G_s \rho_w}{\rho_d} - 1$ | |

Table 13-2. Back Pressure Saturation

| | |
|---|---|
| Cell pressure, $\sigma_3$ | |
| Pore and back pressure, $u$ | |
| Duration (hour) | |
| $B$ value | |
| Piston movement during saturation, $L_1$ (cm) | |

If an additional saturation stage is used, the data should be recorded.

Table 13-3. Consolidation

| | |
|---|---|
| Cell pressure, $\sigma_3$ | |
| Pore and back pressure, $u$ | |
| Effective stress, $\sigma' = \sigma_3 - u$ | |
| Duration (hour) | |
| Piston movement during consolidation, $L_2$ (cm) | |

Table 13-4. Permeability Measurements

Length of specimen, $L = L_0 - L_1 - L_2 =$

Cross-sectional area (from Table 1), $A =$

Temperature, T =

Flow direction:

| | **Test 1** | **Test 2** | **Test 3** | **Test 4** |
|---|---|---|---|---|
| Initial inflow burette reading, $V_0$ | | | | |
| Final inflow burette reading, $V_t$ | | | | |
| Inflow, $V_{in} = V_0 - V_t$ (ml) | | | | |
| Initial outflow burette reading, $V_0$ | | | | |
| Final outflow burette reading, $V_t$ | | | | |
| Time duration, $\Delta t$ (sec) | | | | |
| Outflow, $V_{out} = V_0 - V_t$ (ml) | | | | |
| Average flow volume $Q = (V_{in} + V_{out})/2$ (ml) | | | | |
| Initial pore pressure | | | | |
| Initial back pressure | | | | |
| Final pore pressure | | | | |
| Final back pressure | | | | |
| Initial pressure difference | | | | |
| Initial head loss, $h_1$ | | | | |
| Final pressure difference | | | | |
| Final head loss, $h_2$ | | | | |
| Average head loss, $\Delta h = (h_1 + h_2)/2$ | | | | |
| $k = \dfrac{Q \cdot L}{A \cdot \Delta h \cdot \Delta t}$ (cm/s) | | | | |

## 8. Results and Analyses

(1) Complete the calculations in Tables 13-1, 13-2, 13-3, and 13-4.

(2) Calculate the average of the hydraulic conductivity, $k_{T^0C}$ .

(3) Correct the hydraulic conductivity to that at 20°C (68°F) by multiplying $k_{T^0C}$ by the ratio of the viscosity of water at the test temperature, $\eta_{T^oC}$, to the viscosity of water at 20°C (68°F), $\eta_{20^oC}$.

$$k_{20^0C} = \frac{\eta_{T^0C}}{\eta_{20^0C}} k_{T^0C}$$

Table 13-5. Variation of $\frac{\eta_{T^0C}}{\eta_{20^0C}}$ with Temperature

| **Temperature, T (°C)** | $\frac{\eta_{T^0C}}{\eta_{20^0C}}$ | **Temperature, T (°C)** | $\frac{\eta_{T^0C}}{\eta_{20^0C}}$ |
|---|---|---|---|
| 15 | 1.135 | 23 | 0.931 |
| 16 | 1.106 | 24 | 0.910 |
| 17 | 1.077 | 25 | 0.889 |
| 18 | 1.051 | 26 | 0.869 |
| 19 | 1.025 | 27 | 0.850 |
| 20 | 1.000 | 28 | 0.832 |
| 21 | 0.976 | 29 | 0.814 |
| 22 | 0.953 | 30 | 0.797 |

$\eta_{20^oC}$ =1.0020 mPa·s

## SAMPLE DATA COLLECTION, RESULTS, AND ANALYSES

Table 13-1. Initial Specimen Dimension and Density

| | |
|---|---|
| Initial water content (%) | 9.1 |
| Specific gravity, $G_s$ | 2.65 |
| Length, $L_0$ (cm) | 14.13 |
| Diameter, $D$ (cm) | 7.21 |
| Mass of specimen, $M$ (g) | 1045.9 |
| Cross-sectional area, $A$ (cm$^2$) | 40.83 |
| Volume, $V$ (cm$^3$) | 576.90 |
| Moist density, $\rho$ (g/cm$^3$) | 1.81 |
| Dry density, $\rho_d$ (g/cm$^3$) | 1.66 |
| Void ratio, $e = \frac{G_s \rho_w}{\rho_d} - 1$ | 0.55 |

Table 13-2. Back Pressure Saturation

| | |
|---|---|
| Cell pressure, $\sigma_3$ | 5.0 psi |
| Pore and back pressure, $u$ | 3.0 psi |
| Duration (hour) | 48 |
| $B$ value | 0.95 |
| Piston movement during saturation, $L_1$ (cm) | 0.2 |

If an additional saturation stage is used, the data should be recorded.

Table 13-3. Consolidation

| | |
|---|---|
| Cell pressure, $\sigma_3$ | 13.0 psi |
| Pore and back pressure, $u$ | 3.0 psi |
| Effective stress, $\sigma' = \sigma_3 - u$ | 10.0 psi |
| Duration (hour) | 24 |
| Piston movement during consolidation, $L_2$ (cm) | 0.4 |

Table 13-4. Permeability Measurements

Length of specimen, $L = L_0 - L_1 - L_2$ = 14.13-0.2-0.4 = 14.07cm

Cross-sectional area (from Table 1), $A$ = 40.83cm$^2$

Temperature, T = 21.0°C

Flow direction: Downward

| | Test 1 | Test 2 | Test 3 | Test 4 |
|---|---|---|---|---|
| Initial inflow burette reading, $V_0$ (ml) | 19.0 | 19.0 | 19.0 | 19.0 |
| Final inflow burette reading, $V_t$ (ml) | 3.8 | 5.0 | 3.0 | 4.5 |
| Inflow, $V_{in} = V_0 - V_t$ (ml) | 15.2 | 14.0 | 16.0 | 14.5 |
| Initial outflow burette reading, $V_0$ (ml) | 3.0 | 3.0 | 3.0 | 3.0 |
| Final outflow burette reading, $V_t$ (ml) | 18.1 | 16.9 | 19.0 | 17.5 |
| Time duration, $\Delta t$ (sec) | 480 | 453 | 495 | 466 |
| Outflow, $V_{out} = V_t - V_0$ (ml) | 15.1 | 13.9 | 16.0 | 14.5 |
| Average flow volume $Q = (V_{in} + V_{out})/2$ (ml) | 15.15 | 13.95 | 16.0 | 14.5 |
| Initial pore pressure (psi) | 7.98 | 8.00 | 7.97 | 7.97 |
| Initial back pressure (psi) | 3.00 | 3.01 | 2.98 | 2.97 |
| Final pore pressure (psi) | 7.98 | 8.00 | 7.97 | 7.97 |
| Final back pressure (psi) | 3.00 | 3.01 | 2.98 | 2.98 |
| Initial pressure difference (psi) | 4.98 | 4.99 | 4.99 | 5.00 |
| Initial head loss, $h_1$ (cm) | 350.29 | 351.00 | 351.00 | 351.70 |
| Final pressure difference (psi) | 4.98 | 4.99 | 4.99 | 4.99 |
| Final head loss, $h_2$ (cm) | 350.29 | 351.00 | 351.00 | 351.00 |
| Average head loss, $\Delta h = (h_1 + h_2)/2$ (cm) | 350.29 | 351.00 | 351.00 | 351.35 |
| $k = \dfrac{Q \cdot L}{A \cdot \Delta h \cdot \Delta t}$ (cm/s) | $3.1\times10^{-5}$ | $3.0\times10^{-5}$ | $3.2\times10^{-5}$ | $3.0\times10^{-5}$ |

(1 psi = 70.34 cm water head)

The average of the hydraulic conductivity: $k_{21^0C}$ = $3.1\times10^{-5}$ cm/s

Correct the hydraulic conductivity to that at 20°C (68°F) by multiplying $k_{21^0C}$ by the ratio of the viscosity of water at the test temperature, $\eta_{21^{o}C}$, to the viscosity of water at 20°C (68°F), $\eta_{20^{o}C}$.

$$k_{20^0C} = \frac{\eta_{21^0C}}{\eta_{20^0C}} k_{21^0C} = 0.976\times(3.1\times10^{-5}) = 3.0\times10^{-5} cm/s$$

Name ______________________________

# Test 13: Permeability Test Using Flexible Wall Permeameter

Question 1: What is the main advantage of the flexible wall permeameter over the rigid wall permeameter?

Question 2: What is the definition of *B* value? What is the minimum *B* value at which the soil specimen can be considered saturated?

Question 3: During the consolidation stage, the cell pressure is 20 psi, both the pore and back pressures of the soil specimen are 15 psi, what are the total stress, pore water pressure, and the effective stress of the soil specimen?

# TEST 14

# Triaxial Compression Tests (UU, CU, CD)

## 1. Introduction

The objective of this test is to determine the strength and stress-strain relationships of a cylindrical specimen. The specimen is subjected to a confining pressure in a triaxial cell or chamber. After specimen saturation, the cell pressure is increased to achieve a desired effective stress; drainage is controlled to achieve either consolidated or unconsolidated condition; then a vertical deviator stress is applied to the specimen at a constant rate of axial deformation (strain controlled), and the compression is conducted under drained or undrained condition until specimen failure. Three test conditions, unconsolidated-undrained (UU), consolidated-undrained (CU), and consolidated-drained (CD), can be simulated.

The specimen can be undisturbed or remolded. Any type of soil (sand, silt, clay) can be tested. The test setup is illustrated in Figure 14-1.

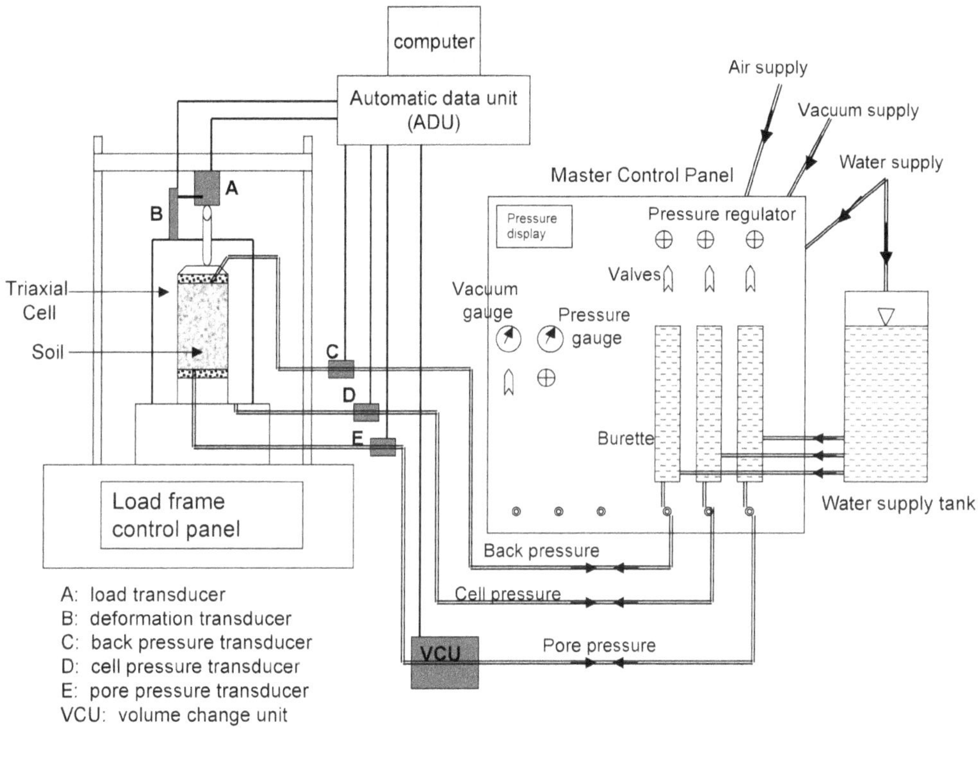


Figure 14-1. Triaxial equipment setup for compression test.

The shear strength of soil, $\tau_f$ , is defined using the Mohr-Coulomb failure criterion:

$$\tau_f = c' + \sigma' \tan\phi'$$

where:

$c'$ = cohesion based on effective stress,

$\phi'$ = internal friction angle based on effective stress, and

$\sigma'$ = effective stress, and

$$\sigma' = \sigma - u$$

where:

$\sigma$ = total stress, and
$u$ = pore water pressure.

In the triaxial compression test, the total stress and pore water pressure are measured using load and pressure transducers.

Mohr's circles that represent the stress conditions of a soil specimen at failure can be plotted as follows:

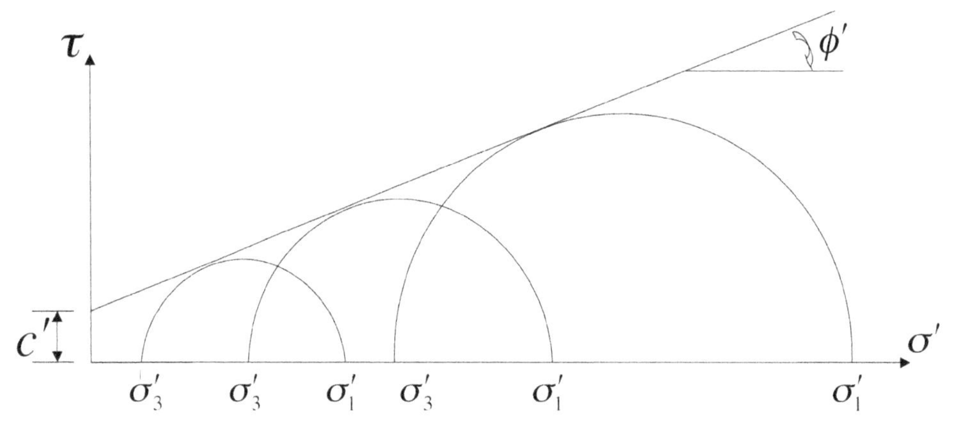


Figure 14-2. Mohr's circles and Mohr-Column failure envelope for CU and CD tests.

For CU and CD tests, specimens that are consolidated under different effective stresses ($\sigma_3'$ ) fail at different vertical effective stresses ($\sigma_1'$ ). Therefore, the Mohr's circles at specimen failure have different dimensions, as shown in Figure 14-2. The tangent line to the Mohr's circles is the Mohr-Coulomb failure envelope, from which, the $c'$ and $\phi'$ can be determined.

In the UU test, effective stress, $\sigma_3'$ , does not increase since drainage is not allowed. Therefore, at failure, the vertical effective stress, $\sigma_1'$ remains the same. So the Mohr's circles that represent total stress conditions at specimen failure have the same diameter, i.e.,

$$\sigma_1 - \sigma_3 = (\sigma_1' + u) - (\sigma_3' + u) = \sigma_1' - \sigma_3' = \text{constant}$$

The failure envelope is a horizontal line, as shown in Figure 14-3.

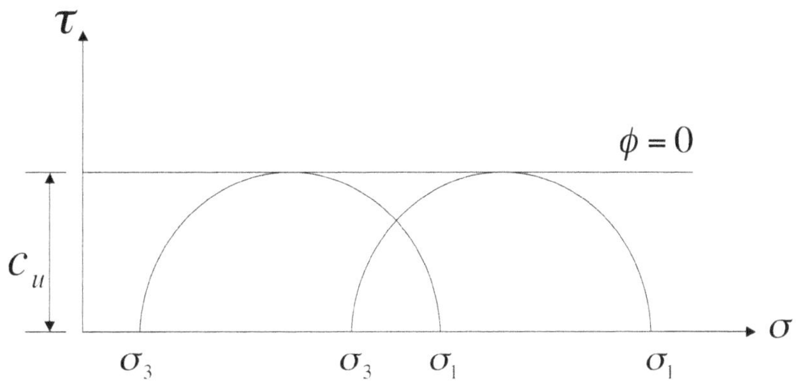


Figure 14-3. Mohr's circles and Mohr-Column failure envelope for the UU test.

## 2. References

- ASTM D 2850-03a "Standard Test Methods for Unconsolidated-Undrained Triaxial Compression Test on Cohesive Soils," *2006 Annual Book of ASTM Standards*, Volume 04.08, Soil and Rock (I). ASTM International.
- ASTM D 4767-04 "Standard Test Methods for Consolidated Undrained Triaxial Compression Test for Cohesive Soils," *2006 Annual Book of ASTM Standards*, Volume 04.08, Soil and Rock (I). ASTM International.

## 3. Soils and Other Materials

- Any type of soil (sand, silt, clay) can be tested
- Tap water

## 4. Apparatus

- Water supply, air supply, and vacuum supply sources
- A master control panel that can supply and regulate pressurized water, pressurized air, and vacuum; the panel should include burettes to measure the volume of influent to and effluent from the specimen
- Axial loading device that is driven by an electric motor to control the strain rate of compression
- A triaxial cell that encases the specimen; the cell should have a rigid and impermeable top cap and a base, a piston, and outlets to the control panel
- Water supply tank that can sustain a pressure up to 150 psi (1034kPa).
- Three water-pressure transducers, one volume-change transducer, one deformation transducer (e.g., LVDT), and one load cell with a data acquisition system or readout
- Flexible membrane whose diameter is 90–95% of that of the specimen, two porous stones whose diameter difference with that of the specimen is ±5%, filter paper, and four o-rings whose diameter is less than 90% of that of the specimen

- Sample compaction equipment, such as tamper and mold of appropriate dimension (not needed if using an undisturbed specimen)
- Sample extruder (not needed if the sample compaction mold is split type)
- Trimming equipment (trimming turntable, wire saw, straightedge)
- Membrane tester (to test for any tears or holes in the membrane)
- Membrane expander used to place membrane onto the specimen
- O-ring placing tool
- Balance
- Water content container
- Drying oven
- Timer
- Caliper
- Thermometer
- Rigid tubings (so they do not expand under pressurized water)

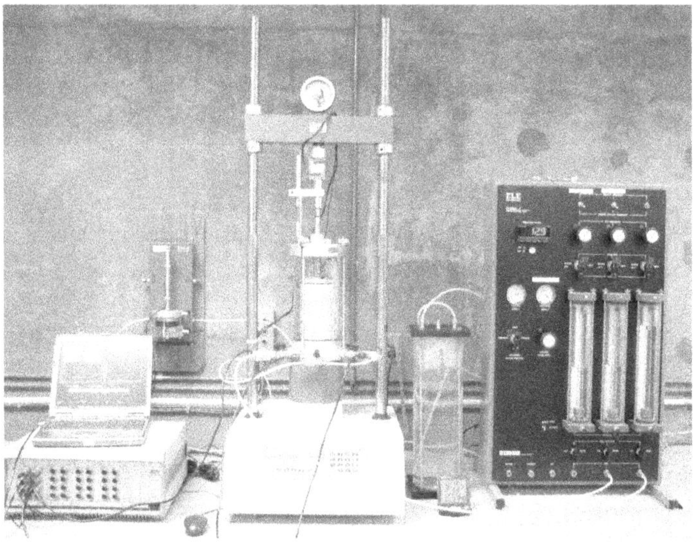

Figure 14-4. Triaxial compression test setup.

## 5. Preparation of Test Specimen

The specimen shall have a minimum diameter of 33 mm (1.3 in). The ratio of the height to diameter should be between 2.0 to 2.5. The largest particle size shall be smaller than 1/6 of the specimen diameter.

**If using undisturbed specimens:** Specimens obtained by tube sampling may be tested without trimming except for cutting the end surfaces.

**If using laboratory-compacted specimens:** The soil should be first moistened at the specified water content and cured for 16 hours if the soil is silt or clay. The soil should be covered to prevent moisture loss. The soil is then compacted in a split mold using several layers (Figure 14-5). Each layer should have the same amount of soil and be compacted to the same thickness to achieve homogeneity. The surface of each previously compacted layer shall be lightly scarified with a straightedge to minimize the effect of strata.

The test specimen shall be mounted immediately in the triaxial cell after trimming to prevent moisture loss.

Figure 14-5. Specimen compaction device.

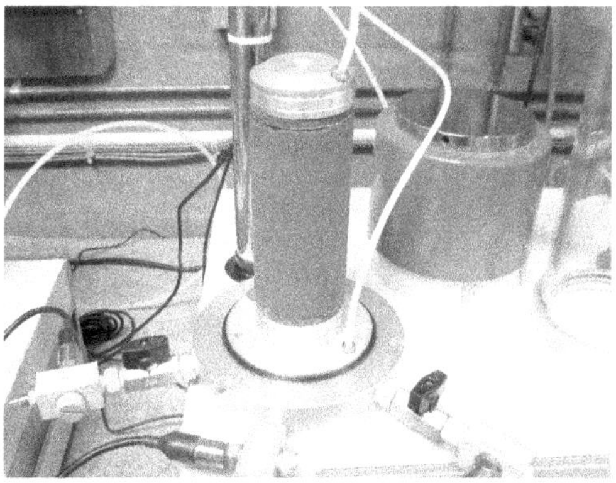

Figure 14-6. Mounting the specimen.

## 6. Test Procedure

(1) Mount the specimen (Figure 14-6):

- Prepare de-aired water in the water supply tank by applying vacuum to the tank for at least 30 minutes. The vacuum should be high enough to initially generate air bubbles in the tank.
- Place the membrane on the membrane tester and supply about 1 psi to inflate the membrane slightly to check for possible leaks.
- After trimming the specimen, measure the diameter, the length, and the mass of the specimen.
- In the bottom-up sequence, place a porous stone, filter paper, specimen, filter paper, porous stone, and top cap on the base plate.
- Place the membrane in the membrane expander and apply vacuum to expand the membrane to adhere to the inside wall of the expander. Then slide the membrane over the specimen, and release the vacuum.
- Place two O-rings to seal the membrane to the base and two O-rings to seal the membrane to the top cap, using an O-ring expander.
- Attach flow tubing(s) to the top cap. The flow tubing connects to the back-pressure port.
- Enclose the specimen with the clear cylinder and the cell top. Tighten the cell top.
- Place the triaxial cell onto the axial loading device. Align the triaxial cell.
- Attach the deformation transducer and the load cell to the cell top. Both transducers should have readouts or be connected to a data acquisition unit.
- Bring the piston into contact with the specimen cap. The piston helps keep the specimen vertical. Record the piston location.

(2) Set up transducers and flow lines:

- Connect the cell-pressure transducer, pore-pressure transducer, and back-pressure transducer to the corresponding ports, which are connected to the control panel via rigid tubing. The transducers should have readouts, or the transducers can be connected to data acquisition units so the readings can be displayed on a computer (appropriate software may be needed).

  **Note:** The pore-pressure transducer should be between a valve and the triaxial cell.

- Connect the volume-change unit between the control panel and back-pressure port.
- Fill the triaxial cell with water and ensure no air bubbles are in the cell.
- Remove any air bubbles in the flow lines by applying a small pressure (1 psi) and opening the water-bleeding ports. The valves to the specimen should be closed during this process. Then reduce the pressure to zero.
- Reset the transducers to zero.
- Close all pressure valves.

(3) Saturation (this step is not needed for the UU test):
When the initial degree of saturation is less than 85%, applying a low vacuum to the specimen may help reduce continuous air in the specimen. When using a vacuum, the top porous stone and filter paper should be dry. The steps are detailed below:

- Empty the burette that connects to the back-pressure valve, and adjust the vacuum in the burette to 35kPa (5 psi).
- Open the back-pressure valve and keep the pore- and cell-pressure valves closed. Vacuum is applied to the specimen from the top. Maintain the vacuum for 10 minutes. Then reduce the vacuum to 20kPa (3 psi).
- Increase the cell pressure to a desired value, such as 70kPa (10 psi).
- Slowly open the cell-pressure valve and wait 2 minutes for stabilization.
- Slowly open the pore-pressure valve. Water will enter the specimen from the bottom under vacuum. If needed, a small pore pressure (e.g., 5 psi) can be applied at the specimen bottom to facilitate saturation; if so, the cell pressure should be at least 1 psi more than the pore pressure to prevent membrane bloating.
- When water appears in the burette connected to the top of the specimen, slowly reduce the vacuum to zero.
- Close both the pore- and back-pressure valves, and fill de-aired water in the burette that connects to the back-pressure outlet.
- Adjust both the pore and back pressures. The pore and back pressures should be the same. The difference between the cell pressure and the pore/back pressure should be between 15 to 35kPa (2 to 5 psi).
- Open the pore- and back-pressure valves slowly and simultaneously. Water will enter into the specimen from the top and the bottom. The air will dissolve into the de-aired water. This process may take 24 hours to several days, depending on the soil.
- Check $B$ value to verify saturation:

$$B = \frac{\Delta u}{\Delta \sigma_3}$$

where:

$\Delta\sigma_3$ = cell pressure increase,

$\Delta u$ = pore pressure increase.

First close the cell-, pore-, and back-pressure valves; then increase the cell pressure by 70kPa (10 psi), and open the cell-pressure valve. Wait for 2 minutes to stabilize. Read the pore-pressure reading and calculate the pore-pressure increase. The $B$ value can then be calculated. A specimen can be considered saturated if $B \geq 0.95$. Otherwise, increase the pore and back pressure by 10 psi (same as $\Delta\sigma_3$); then open both the back- and pore- pressure valves. Wait for another 24–48 hours, and check the $B$ value again. Repeat this process until $B \geq 0.95$.

(4) Consolidation (This step is not needed for the UU test):
- Bring the piston into contact with the specimen and record the piston location.
- Close the cell-, pore-, and back-pressure valves. Adjust the cell pressure to achieve desired effective stress, which is the cell pressure minus the pore pressure.
- Open the cell-pressure valve and wait 2 minutes for the pressure to stabilize. Then open both the pore- and back-pressure valves to start consolidation. Water should slowly drain into both the pore- and back-pressure burettes.
- Wait 24 hours for consolidation to complete. At the end of the consolidation, bring the piston into contact with the specimen cap. Record the piston location and calculate the piston movement so that the specimen height at the end of the consolidation can be calculated.

(5) Axial loading:
- For the UU and CU tests: close both the pore- and back-pressure valves. Select an axial strain rate for the loading (1%/min for plastic materials and 0.3%/min for brittle materials).
- For the CD test: keep the pore- and back-pressure valves open. Select an axial strain rate that is slow enough to permit the drainage and the dissipation of excessive pore water pressure.
- Start the axial loading. The deviator axial load is measured by the load cell. The axial deformation is measured by the LVDT transducer. In the UU and CU tests, pore pressure increases due to the undrained loading, and it is measured by the pore-pressure transducer. In the CD test, pore pressure is monitored to ensure the dissipation of excessive pore pressure. Take sufficient readings to define the stress-strain curve. At the axial strain rate specified for the UU or CU test, the elapsed time to reach maximum deviator stress is approximately 15 to 20 minutes.
- Continue loading to 15% axial strain. Loading may be stopped when the deviator stress has peaked then dropped 20% or the axial strain has reached 5% beyond the strain at which the peak in deviator occurred.
- After completion of the test, remove the axial load by lowering the specimen.
- Release the pore and back pressure and the cell pressure simultaneously. Note: Pore/back pressure should not exceed cell pressure to prevent membrane bloating. Then drain the water in the cell and carefully remove the specimen from the cell.
- Carefully remove the membrane. Sketch a picture or take a photograph of the specimen showing the failure surface.
- Determine the water content of the specimen. Free water remaining on the specimen surface should be blotted away before the water content measurement.

(6) For the CU and CD tests, redo the entire test at a higher effective stress to obtain at least two additional Mohr's circles to define the Mohr-Coulomb failure envelope.

## 7. Data Collection and Calculation

Table 14-1. Initial Specimen Dimension and Density

| | |
|---|---|
| Initial water content (%) | |
| Specific gravity of soil solids, $G_s$ | |
| Specimen length, $L_0$ (cm) | |
| Specimen diameter, $D$ (cm) | |
| Mass of specimen, $M$ (g) | |
| Cross-sectional area, $A$ (cm$^2$) | |
| Volume, $V$ (cm$^3$) | |
| Moist density, $\rho$ (g/cm$^3$) | |
| Dry density, $\rho_d$ (g/cm$^3$) | |
| Void ratio, $e = \frac{G_s \rho_w}{\rho_d} - 1$ | |

Table 14-2. Back Pressure Saturation (for CU and CD Tests)

| | |
|---|---|
| Cell pressure, $\sigma_3$ | |
| Pore and back pressure, $u$ | |
| Duration (hour) | |
| $B$ value | |
| Piston movement during saturation, $L_1$ (cm) | |

If an additional saturation stage is used, the data should be recorded.

Table 14-3. Consolidation (for CU and CD Tests)

| | |
|---|---|
| Cell pressure, $\sigma_3$ | |
| Pore and back pressure, $u$ | |
| Effective stress, $\sigma' = \sigma_3 - u$ | |
| Duration (hour) | |
| Piston movement during consolidation, $L_2$ (cm) | |

Table 14-4. Axial Loading

Length of specimen before loading, $L = L_0 - L_1 - L_2 =$
Cross-sectional area (from Table 1), $A =$
Specified strain rate =

| | |
|---|---|
| Strain at the maximum deviator stress, $\varepsilon$ (%) | |
| Maximum deviator stress, $\Delta\sigma_d$ | |
| Cell pressure, $\sigma_3$ | |
| Total vertical stress, $\sigma_1 = \sigma_3 + \Delta\sigma_d$ | |
| Pore water pressure at failure, $u$ | |
| Effective minor principal stress, $\sigma_3' = \sigma_3 - u$ | |
| Effective major principal stress, $\sigma_1' = \sigma_1 - u$ | |

## 8. Results and Analyses

(1) Plot the stress vs. strain and pore pressure vs. strain relationship curves during axial loading.
(2) Complete the calculations in Tables 14-1, 14-2, 14-3, and 14-4.
(3) Plot the Mohr's circles and obtain the Mohr-Coulomb failure envelope.
(4) Determine the (effective) cohesion and (effective) internal friction angle.

# SAMPLE DATA COLLECTION, RESULTS, AND ANALYSES

In this sample test report, Load 50® triaxial setup equipped with DS7 software (ELE International) was used. Consolidated Undrained (CU) compression tests were conducted on three remolded silty sand specimens under different confining pressures. The group test report for the three specimens follows.

**Consolidated Undrained Triaxial Compression Test**
**with measurement of Pore Pressure**

ELE International

| | | | |
|---|---|---|---|
| **Client** | | **Lab Ref** | |
| **Project** | | **Job** | |
| **Borehole** | | **Sample** | 1 |

| Test Details | | | |
|---|---|---|---|
| Standard | | Particle Density | 2.65 Mg/m3 (Assumed) |
| Test Definition | Consolidated Undrained | Drainage Location | Top |
| Sample Type | | Lab. Temperature | 20.0 deg.C |
| Sample Description | Remolded silty sand | | |
| Variations from Procedure | None | | |

| Specimen Details | | | |
|---|---|---|---|
| Specimen Reference | A | Description | |
| Depth within Sample | 0.00 mm | Orientation within Sample | |
| Initial Height | 76.00 mm | Initial Diameter | 38.00 mm |
| Preparation | | Moisture Content | 13.5 % (trimmings: 13.5 %) |
| Bulk Density | 1.85 Mg/m3 | Membrane Thickness | 0.30 mm |
| Comments | | | |

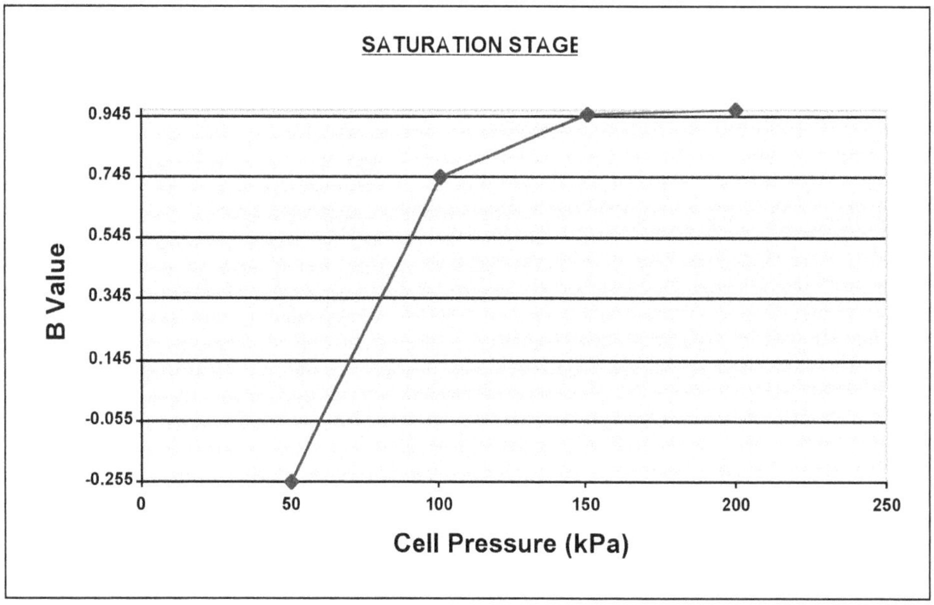


| | | | |
|---|---|---|---|
| **Saturation Method** | Back Pressure Increments | **Cell Increments** | 50.3,50.4,49.6,49.5,kPa |
| **Final Cell Pressure** | 199.8kPa | **Back Increments** | 45.1,50.9,49.3,kPa |
| **Final Pore Pressure** | 194.4kPa | **Final B Value** | 0.967 |

**Consolidated Undrained Triaxial Compression Test**
**with measurement of Pore Pressure**

| | | | |
|---|---|---|---|
| **Client** | | **Lab Ref** | |
| **Project** | | **Job** | |
| **Borehole** | | **Sample** | 1 |

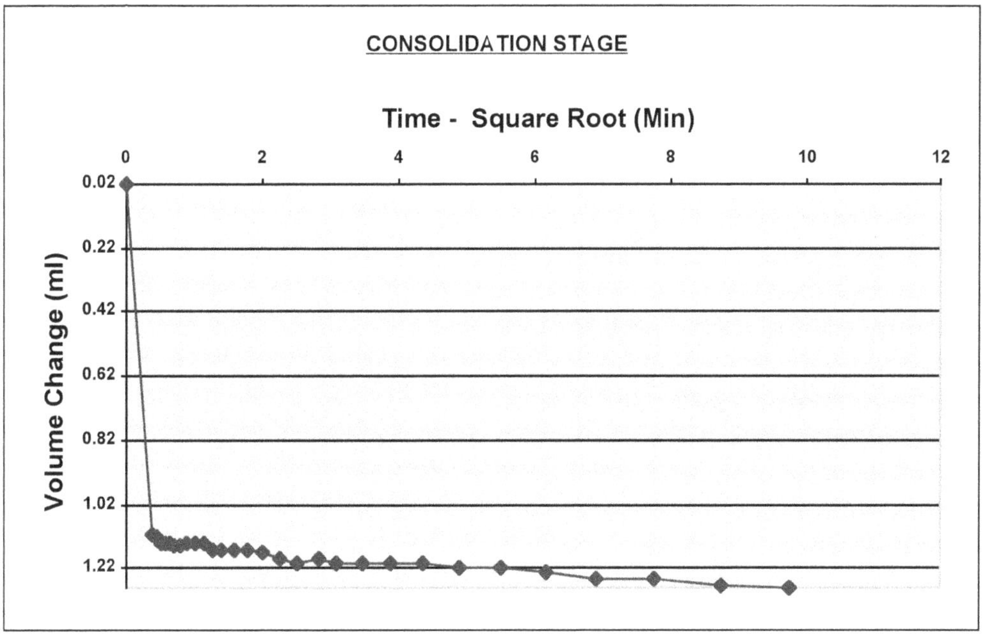


| | | | |
|---|---|---|---|
| **Cell Pressure** | 218.7kPa | **Back Pressure** | 144.0kPa |
| **Effective Pressure** | 74.7kPa | **Final Pore Pressure Dissipation** | 97.54% |
| **Final Pore Pressure** | 145.6 kPa | | |

**Consolidated Undrained Triaxial Compression Test**
**with measurement of Pore Pressure**

| | | | |
|---|---|---|---|
| **Client** | | **Lab Ref** | |
| **Project** | | **Job** | |
| **Borehole** | | **Sample** | 1 |

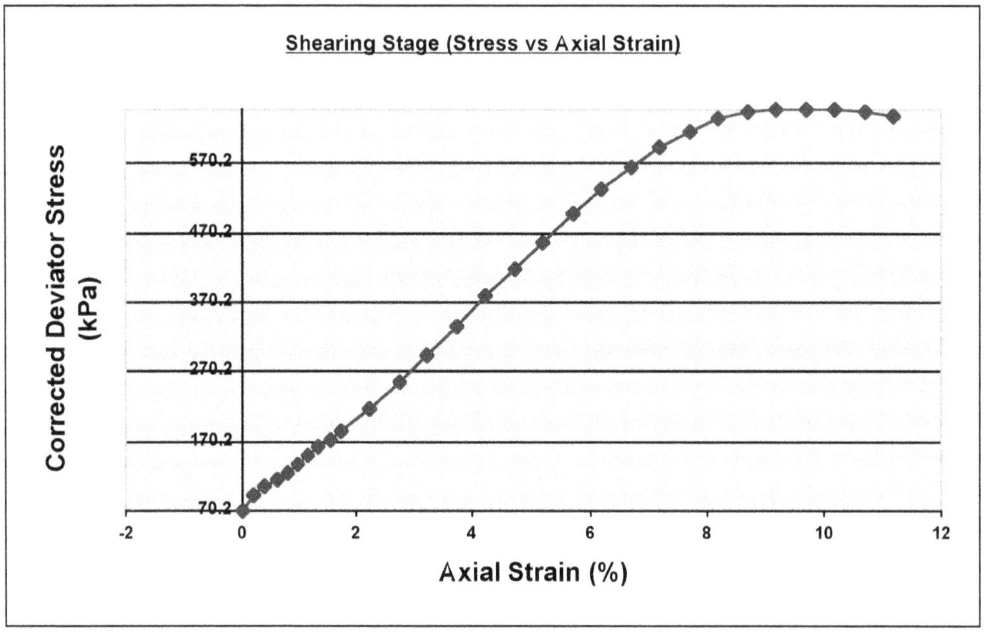

## Consolidated Undrained Triaxial Compression Test
### with measurement of Pore Pressure

| | | | |
|---|---|---|---|
| **Client** | | **Lab Ref** | |
| **Project** | | **Job** | |
| **Borehole** | | **Sample** | 1 |

| Shear Conditions | | | |
|---|---|---|---|
| **Rate of Axial Displacement** | 0.1000mm/min | **Cell Pressure** | 218.7kPa |
| **Initial Pore Pressure** | 174.3kPa | **Effective Stress at Start of Stage** | 44.4kPa |

| Conditions at Failure | | | |
|---|---|---|---|
| **Failure Criterion** | Maximum Deviator Stress | | |
| **Pore Pressure** | -0.7kPa | **Minor Effective Principal Stress** | 218.7kPa |
| **Deviator Stress** | 647.4kPa | **Major Effective Principal Stress** | 866.1kPa |
| **Axial Strain** | 9.20% | **Effective Principal Stress Ratio** | 3.961 |
| **Deviator Stress Correction** | 1.8kPa | | |
| **Density** | 1.91 Mg/m3 | **Moisture Content** | 15.5 % |

| | |
|---|---|
| Tested By and Date: | |
| Checked By and Date: | |
| Approved By and Date: | |

Mode of Failure

# Consolidated Undrained Triaxial Compression Test
## with measurement of Pore Pressure

| | | | |
|---|---|---|---|
| **Client** | | **Lab Ref** | |
| **Project** | | **Job** | |
| **Borehole** | | **Sample** | 2 |

| Test Details | | | |
|---|---|---|---|
| **Standard** | | **Particle Density** | 2.65 Mg/m3 (Assumed) |
| **Test Definition** | Consolidated Undrained | **Drainage Location** | Top |
| **Sample Type** | Remolded silty sand | **Lab. Temperature** | 20.0 deg.C |
| **Sample Description** | | | |
| **Variations from Procedure** | None | | |

| Specimen Details | | | |
|---|---|---|---|
| **Specimen Reference** | B | **Description** | |
| **Depth within Sample** | 0.00 mm | **Orientation within Sample** | |
| **Initial Height** | 76.00 mm | **Initial Diameter** | 38.00 mm |
| **Preparation** | | **Moisture Content** | 14.0 % (trimmings: 14.3 %) |
| **Bulk Density** | 1.80 Mg/m3 | **Membrane Thickness** | 0.30 mm |
| **Comments** | | | |

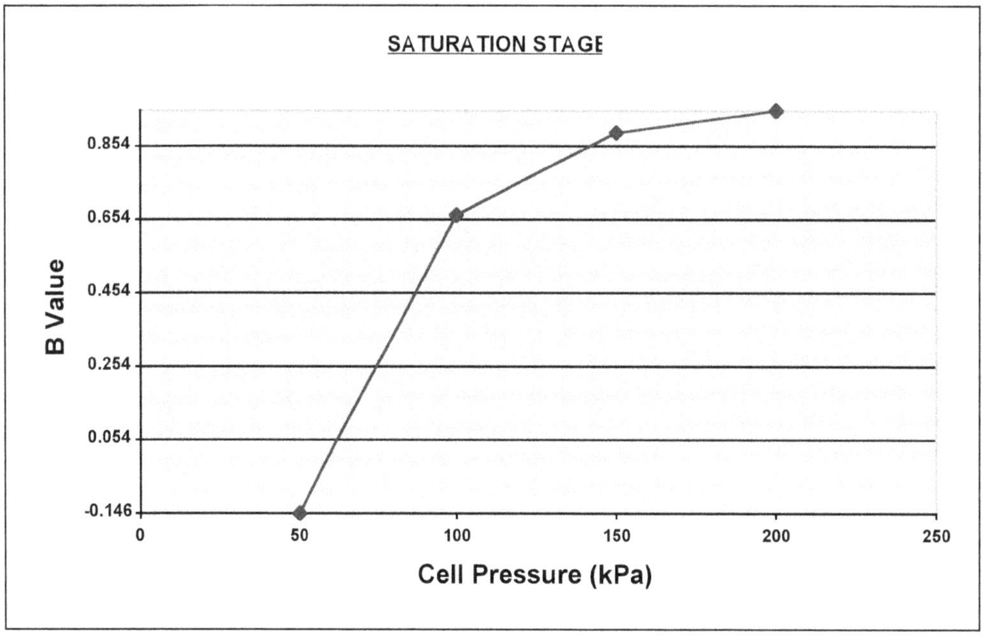


| | | | |
|---|---|---|---|
| **Saturation Method** | Back Pressure Increments | **Cell Increments** | 50.4,49.9,49.9,49.9,kPa |
| **Final Cell Pressure** | 200.0kPa | **Back Increments** | 45.3,50.0,50.2,kPa |
| **Final Pore Pressure** | 192.1kPa | **Final B Value** | 0.948 |

**Consolidated Undrained Triaxial Compression Test**
**with measurement of Pore Pressure**

| | | | |
|---|---|---|---|
| **Client** | | **Lab Ref** | |
| **Project** | | **Job** | |
| **Borehole** | | **Sample** | 2 |

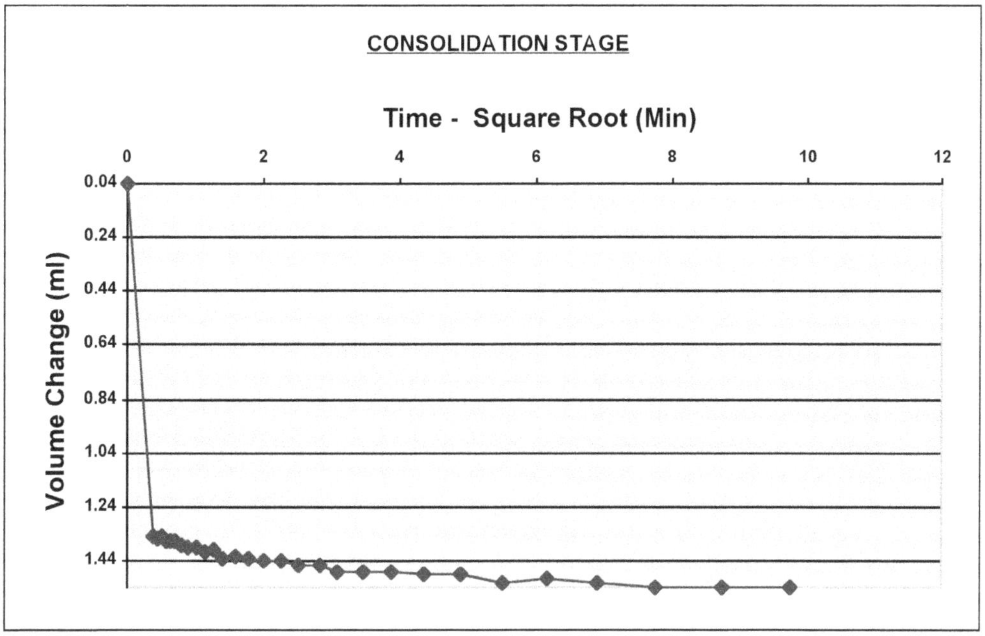


| | | | |
|---|---|---|---|
| **Cell Pressure** | 295.2kPa | **Back Pressure** | 144.9kPa |
| **Effective Pressure** | 150.3kPa | **Final Pore Pressure Dissipation** | 100.44% |
| **Final Pore Pressure** | 144.3 kPa | | |

**Consolidated Undrained Triaxial Compression Test**
**with measurement of Pore Pressure**

| Client | | Lab Ref | |
|---|---|---|---|
| Project | | Job | |
| Borehole | | Sample | 2 |

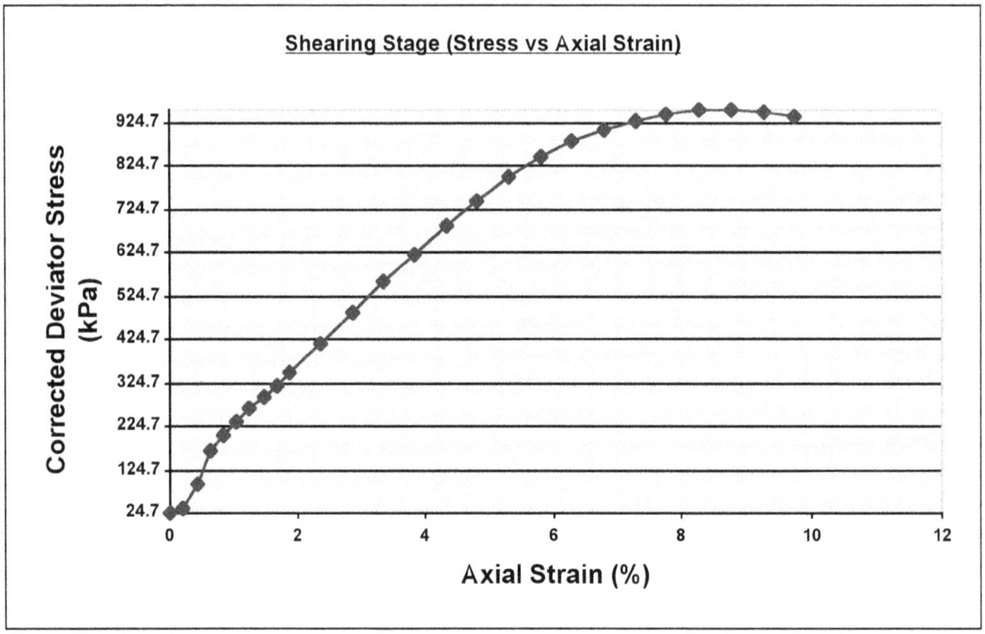

## Consolidated Undrained Triaxial Compression Test
### with measurement of Pore Pressure

| | | | |
|---|---|---|---|
| **Client** | | **Lab Ref** | |
| **Project** | | **Job** | |
| **Borehole** | | **Sample** | 2 |

| Shear Conditions | | | |
|---|---|---|---|
| **Rate of Axial Displacement** | 0.1000mm/min | **Cell Pressure** | 294.5kPa |
| **Initial Pore Pressure** | 156.5kPa | **Effective Stress at Start of Stage** | 138.0kPa |

| Conditions at Failure | | | |
|---|---|---|---|
| **Failure Criterion** | Maximum Deviator Stress | | |
| **Pore Pressure** | -12.9kPa | **Minor Effective Principal Stress** | 306.9kPa |
| **Deviator Stress** | 954.1kPa | **Major Effective Principal Stress** | 1261.1kPa |
| **Axial Strain** | 8.76% | **Effective Principal Stress Ratio** | 4.108 |
| **Deviator Stress Correction** | 1.7kPa | | |
| **Density** | 1.92 Mg/m3 | **Moisture Content** | 19.1 % |

| | |
|---|---|
| Tested By and Date: | |
| Checked By and Date: | |
| Approved By and Date: | |

## Consolidated Undrained Triaxial Compression Test
### with measurement of Pore Pressure

ELE International

| Client | | Lab Ref | |
|---|---|---|---|
| Project | | Job | |
| Borehole | | Sample | 3 |

| Test Details | | | |
|---|---|---|---|
| Standard | | Particle Density | 2.65 Mg/m3 (Assumed) |
| Test Definition | Consolidated Undrained | Drainage Location | Top |
| Sample Type | Remolded silty sand | Lab. Temperature | 20.0 deg.C |
| Sample Description | | | |
| Variations from Procedure | None | | |

| Specimen Details | | | |
|---|---|---|---|
| Specimen Reference | C | Description | |
| Depth within Sample | 0.00 mm | Orientation within Sample | |
| Initial Height | 76.00 mm | Initial Diameter | 38.00 mm |
| Preparation | | Moisture Content | 14.1 % (trimmings: 14.2 %) |
| Bulk Density | 1.86 Mg/m3 | Membrane Thickness | 0.30 mm |
| Comments | | | |

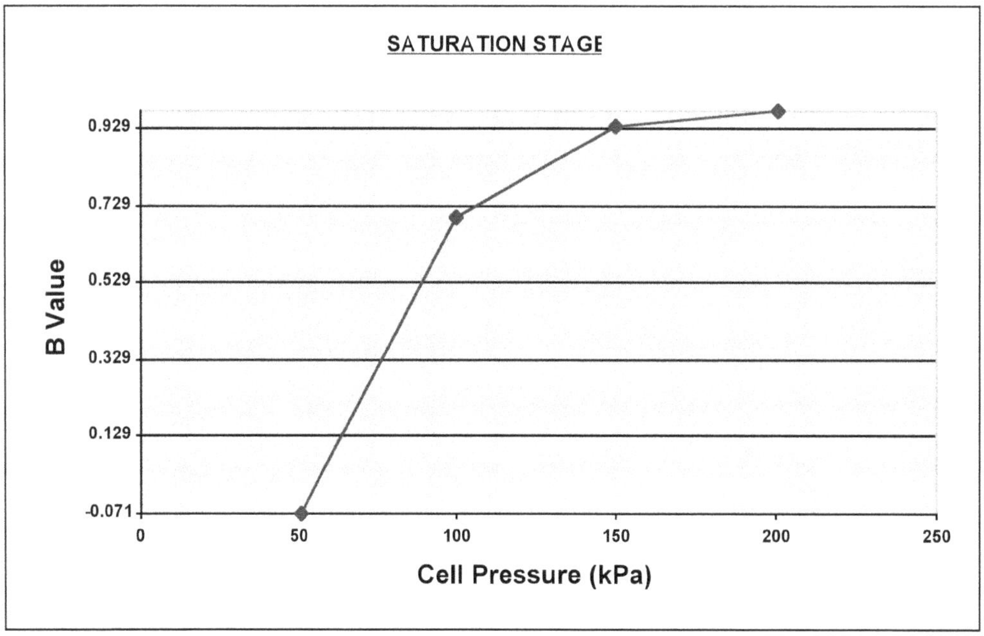


| Saturation Method | Back Pressure Increments | Cell Increments | 50.4,49.9,49.9,51.0,kPa |
|---|---|---|---|
| Final Cell Pressure | 200.6kPa | Back Increments | 44.5,50.5,50.1,kPa |
| Final Pore Pressure | 195.5kPa | Final B Value | 0.975 |

## Consolidated Undrained Triaxial Compression Test
### with measurement of Pore Pressure

| Client | | Lab Ref | |
|---|---|---|---|
| Project | | Job | |
| Borehole | | Sample | 3 |

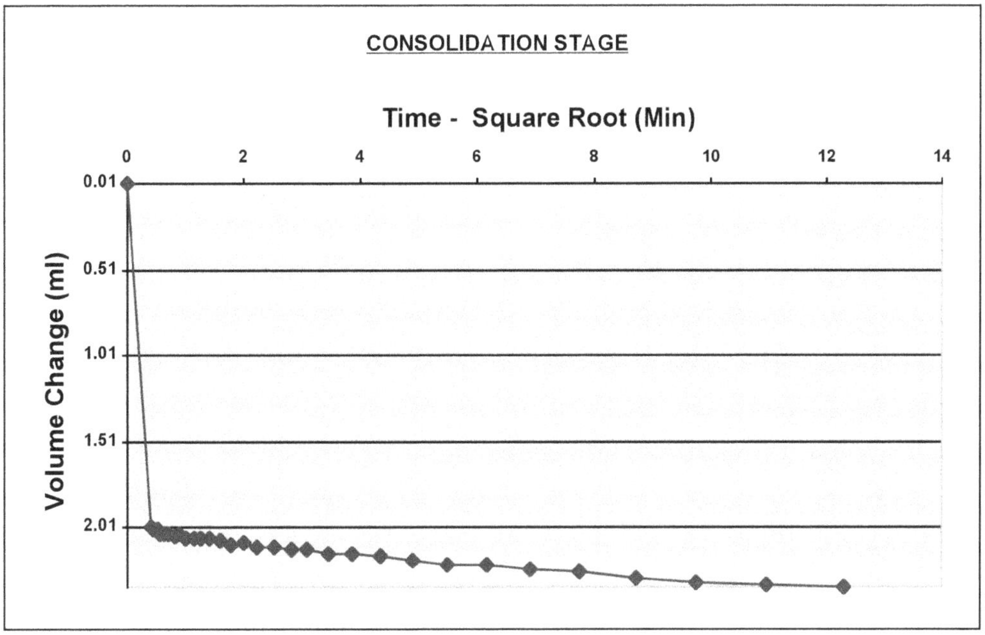


| | | | |
|---|---|---|---|
| **Cell Pressure** | 444.1kPa | **Back Pressure** | 144.3kPa |
| **Effective Pressure** | 299.8kPa | **Final Pore Pressure Dissipation** | 99.77% |
| **Final Pore Pressure** | 145.0 kPa | | |

## Consolidated Undrained Triaxial Compression Test
### with measurement of Pore Pressure

| Client | | Lab Ref | |
|---|---|---|---|
| Project | | Job | |
| Borehole | | Sample | 3 |

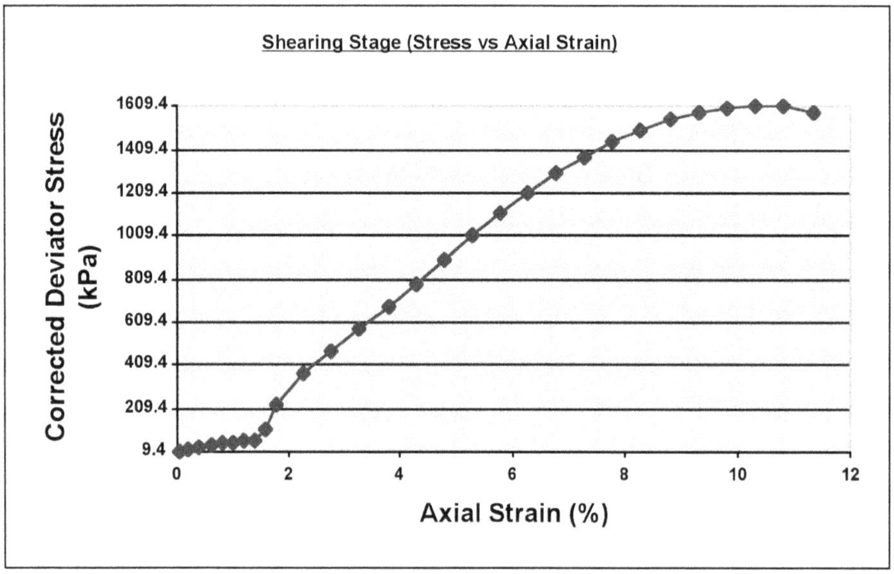


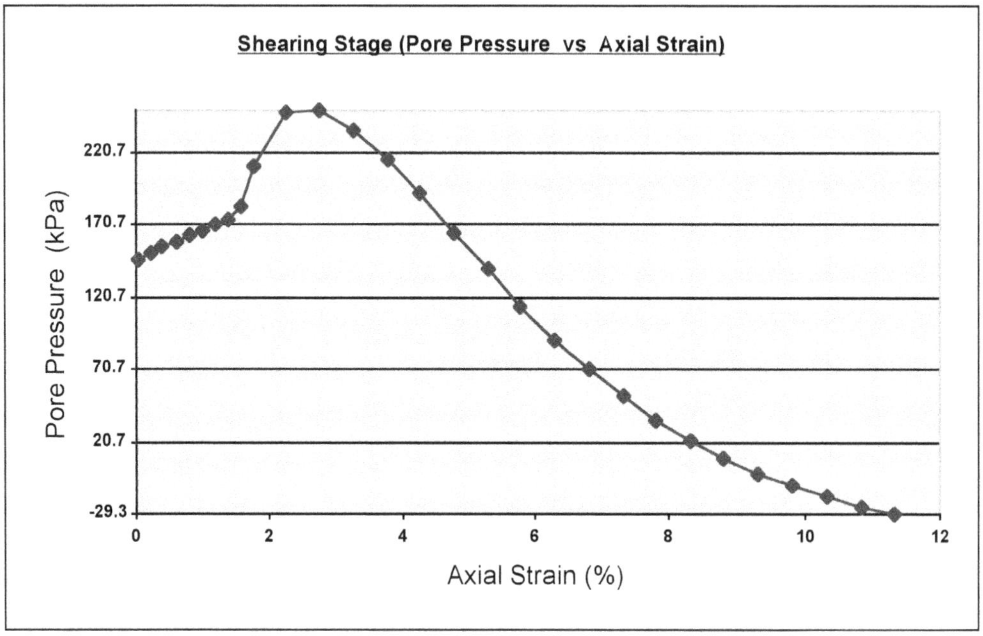

## Consolidated Undrained Triaxial Compression Test
### with measurement of Pore Pressure

| | | | |
|---|---|---|---|
| **Client** | | **Lab Ref** | |
| **Project** | | **Job** | |
| **Borehole** | | **Sample** | 3 |

| Shear Conditions | | | |
|---|---|---|---|
| **Rate of Axial Displacement** | 0.1000mm/min | **Cell Pressure** | 443.1kPa |
| **Initial Pore Pressure** | 146.9kPa | **Effective Stress at Start of Stage** | 296.2kPa |

| Conditions at Failure | | | |
|---|---|---|---|
| **Failure Criterion** | Maximum Deviator Stress | | |
| **Pore Pressure** | -17.7kPa | **Minor Effective Principal Stress** | 460.6kPa |
| **Deviator Stress** | 1611.5kPa | **Major Effective Principal Stress** | 2072.0kPa |
| **Axial Strain** | 10.32% | **Effective Principal Stress Ratio** | 4.499 |
| **Deviator Stress Correction** | 2.0kPa | | |
| **Density** | 1.94 Mg/m3 | **Moisture Content** | 15.4 % |

| | |
|---|---|
| Tested By and Date: | |
| Checked By and Date: | |
| Approved By and Date: | |

Mode of Failure

## Consolidated Undrained Triaxial Compression Test
### with measurement of Pore Pressure

**SUMMARY**

| Test Details | |
|---|---|
| Standard | |

| Specimen Details | | |
|---|---|---|
| Specimen Reference | Effective Minor Principal Stress ($\sigma_3'$) | Effective Major Principal Stress ($\sigma_1'$) |
| A | 218.7kPa | 866.1kPa |
| B | 306.9kPa | 1261.1kPa |
| C | 460.6kPa | 2072.0kPa |
| | | |
| | | |

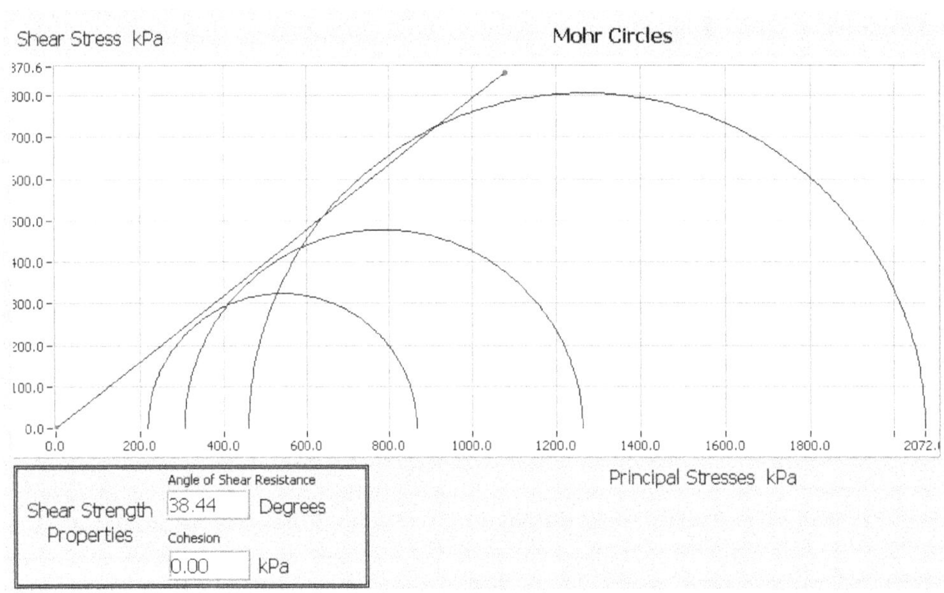

Name ______________________________

# Test 14: Triaxial Compression Tests (UU, CU, CD)

Question 1: Soil's shear strength depends on two soil strength parameters. What are they?

Question 2: What is the definition of B value? What is the minimum B value at which the soil specimen can be considered saturated?

Question 3: A CU triaxial test is conducted on a normally consolidated clay. During the consolidation phase, the cell pressure is 40 psi and the pore water pressure is 10 psi; at the compression failure, the deviator stress is 20 psi, and the compression caused a pore water pressure increase of 20 psi at failure. What are the effective major principal stress ($\sigma_1$?) and effective minor principal stress at failure ($\sigma_3$?)?

Question 4: A CD triaxial test is conducted on a normally consolidated clay. The soil is consolidated at a cell pressure of 40 psi, and the deviator stress to cause failure is 100 psi. At failure, the pore water pressure is 10 psi. What are the effective major principal stress ($\sigma$1?) and effective minor principal stress at failure ($\sigma$3?)?

www.ingramcontent.com/pod-product-compliance
Ingram Content Group UK Ltd.
Pitfield, Milton Keynes, MK11 3LW, UK
UKHW051330130926
15127UKWH00013B/97

9 781524 925888